MATH WITHOUT FEAR

BY DR. CAROL GLORIA CRAWFORD

NEW VIEWPOINTS/VISION BOOKS
A Division of Franklin Watts/New York/London

To my parents
Harry and Gloria Crawford
for their constant encouragement and guidance

ACKNOWLEDGMENTS

I want to express my thanks to Georgetown University and in particular to Elizabeth Beall, Associate Dean for Continuing Education, for the opportunity to create and direct the Math Without Fear program.

Also I wish to express my gratitude to Elza Teresa Dinwiddie, my editor. Without her assistance this book would not have been published at this time.

Selections p. 150–160 adopted from *Mathematics: An Everyday Language* by R.E. Wheeler and E.R. Wheeler © 1979 by John Wiley and Sons, Inc. Reprinted by permission of John Wiley and Sons, Inc.

Selections p. 198–203 adopted from *Practical Algebra* by Peter Selby © 1974 by John Wiley and Sons, Inc. Reprinted by permission of John Wiley and Sons, Inc.

Selections pp. 80–83 and 99–100 from *Basic Algebra: A Semi-Programmed Approach* by L. Gilligan and R. Nenno © 1977 by Goodyear Publishing, Inc. Reprinted by permission of Goodyear Publishing Co.

Selections p. 154 from *Modern Elementary Mathematics*. Second Edition by Malcolm Graham © 1975 by Harcourt Brace Jovanovich, Inc. Reprinted by permission of Harcourt Brace Jovanovich, Inc.

Crawford, Carol Gloria.
Math without fear.

Bibliography: p.
Includes index.
1. Mathematics—1961–   I. Title.
QA39.2.C72   510   80–15353
ISBN 0–531–06377–1
ISBN 0–531–06755–6 (pbk.)

New Viewpoints/Vision Books
A Division of Franklin Watts
730 Fifth Avenue
New York, New York 10019

TABLE OF CONTENTS

FOREWORD

Women have been told for eons that girls have verbal skills and boys have mathematical skills, so obviously there was no point in struggling with math beyond the absolute bare minimum. There are subtle and not so subtle societal pressures that tend to put women into a mold marked "Handle with care—cannot deal with abstract or quantitative concepts—essentially subjective, emotional, and irrational." Unfortunately, this stereotype is a limiting factor in many women's lives.

I can certainly empathize with those who are suffering from "math anxiety," for I grew up with a father who was a college mathematics professor and was considered to be a mathematical whiz. For him it was infinitely easier to *do* my math problems than to *explain* them, and to be absolutely honest, I found that if he didn't do them, a few judicially shed tears facilitated the whole process immensely! I slid through my younger years convinced that I had a mental block against math and that someone would always perform these magical functions for me!

Many years later I find myself working as a university administrator, developing and managing a major continuing education program for adults. In the process of counseling and working with women who are preparing to move back into the mainstream of the work force or academia, I find that it is often evident that they are setting their goals much lower than necessary by underestimating their skills and abilities. One of the reasons is the frequently heard complaint that they can't cope with mathematical concepts, and therefore, they were limiting themselves to careers or degrees that would demand a minimum of computational expertise. Thus they were effectively limiting their life goals and objectives to fit what they believed to be a mental block against math.

As we sought to design courses, workshops, and seminars that would assist women with their reentry problems, it became more and more evident that a vital building block in reactivating their sense of confidence was to help them get over their deeply ingrained fear of dealing with mathematical concepts. As we started our search for someone to teach a course designed to assist women in overcoming "mathophobia" we were referred to Carol Crawford, who was at that time a Ph.D. candidate at Georgetown University. She was recommended to us as a young woman who understood the psychological hang-ups with which many women struggled and who also had a nonthreatening teaching style. Her department chairperson, Dr. John Lagnese, assured us of her professional competence and a subsequent personal interview ended our search for a warm, friendly individual who would launch a math program for women that we had been casually referring to as "Math Without Fear."

Dr. Crawford was at once excited about the possibility of designing such a course and enthusiastically went to work. It is interesting to note that her colleagues in the math department were somewhat amused at her enthusiasm for designing a nonthreatening math program for adults. Unfortunately, all too often those whose lives are dedicated to a particular discipline find it exceedingly difficult, if not impossible, to understand and relate to the emotional trauma of those who have been conditioned to believe they cannot cope with those particular concepts.

The experiences that led us to believe that such a course would be valuable were quickly validated by the strong response to the first offering of Math Without Fear, through the Women's Center of our continuing education program at Georgetown. Registration was so heavy that it was necessary to divide the course into two sections.

One of the exciting aspects of continuing education is the response we get from adults who have benefited in a very positive way from one of our programs. It is not at all unusual to have someone tell us that a certain course or workshop has "changed my life." It is this kind of response that has convinced us that lifelong learning is the cutting edge of education for the future! The response to Carol Crawford's course was especially gratifying and prompted us to add an advanced section for those who were asking for a more sophisticated level.

As with many of our programs, Math Without Fear went through a continual metamorphosis as Dr. Crawford remained sensitive to the needs of her students and incorporated the real life experiences that they brought with them into her classroom material.

It quickly became evident to the students that here was a teacher who cared. The atmosphere in this math class was neither threatening nor competitive, but was structured to meet the needs of the students *where they were*. Carol Crawford shared the discovery of most of our continuing education faculty: that teaching adults is a mutual learning situation, for the students bring to the classroom a wealth of experience. Much of an adult's self-identity is derived from his or her life experience. To ignore or reject such experience is to devaluate that individual, and little learning takes place in such an atmosphere. As you read this book you will be impressed by the way Dr. Crawford has incorporated these life experiences into the curriculum.

As we enter the 1980s, it is gratifying indeed to be involved in the administration of a continuing education program in which we are privileged to be enablers for thousands of adults a year as they pursue the fulfillment of their highest need—that of self-actualization. As a spinoff of this program and through the medium of this book, *Math Without Fear*, Dr. Crawford is able to reach out infinitely farther and share her experience and expertise with you . . . !

It is my sincere hope that you will find within this gem of a book the key to unlock your own sense of self-confidence in dealing with the myriad of mathematical concepts that you encounter on a daily basis and that this sense of being in control will lead you to explore new fields of endeavor so that you may continue to develop to your fullest potential.

Elizabeth R. Beall
Associate Dean
School of Continuing Education
Georgetown University
Washington, D.C. 20057

GLOSSARY

Angle figure formed by the union of two half rays

Area measure of the region enclosed by a geometric figure

Arithmetic mean average found by dividing the sum of a series of numbers by the total number of values being considered

Base a number that is being raised to a power, e.g., in x^2, x is the base

Binomial polynomial with two terms

Central tendency measures the numerical values that are located in the middle of a set of data; extends the idea of average

Congruent figures figures with the same size and shape

Dependent events events that influence each other

Distributive property $a(b + c) = ab + ac$

Equation algebraic sentence stating that two quantities are equal

Equivalent equations equations with exactly the same solutions

Event subcollection of outcomes corresponding to an experiment

Exponent power of a number or variable, e.g., in x^2, 2 is the exponent

Factor one of two or more quantities that, when multiplied, give another quantity

Factoring process of writing a polynomial as a product of terms

Frequency number of times a particular piece of data appears

Frequency distribution method to organize large amounts of data

Histogram graph using vertical bars to display a frequency distribution

Independent events events that have no influence on each other

Like terms terms that are exactly the same except for numerical coefficients

Line graph graph using line segments to display a frequency

Line collection of points with no width, no thickness, and infinite length

Line segment, $\overline{AB}$ set of points between two points, *A* and *B*, on a given line

Literal number another name for a variable

Mathematical expectation payoff, or gain, corresponding to an outcome; equal to the probability of the outcome times the amount to be gained

Median middle value of a set of numbers arranged in increasing order

Mode most common value in a set of numbers

Monomial algebraic expression consisting of one term

Multiplication principle counts the number of ways to do an experiment consisting of a series of steps in which each step can be done in one or more ways

Normal distribution bell-shaped distribution so that the mean, median, and mode all coincide

Numerical coefficient number part of an algebraic term, e.g., in $3xyz$, 3 is the numerical coefficient

Odds ratio of number of ways outcome is favorable to number of ways outcome is not favorable

Outcome result of an experiment

Pentagon polygon with five sides

Percent means per hundred; a fraction whose denominator is 100

Percentage results obtained by taking a certain percent of a number

Perimeter distance around a geometric figure

Permutation arrangement of objects

Plane flat surface that continues lengthwise and sideways without end

Plane curve set of points in a plane that can be drawn without lifting your pencil from the paper

Polygon simple closed plane curve formed by three or more line segments

Polynomial algebraic expression with two or more terms; each term is a product of letters and/or numbers

Population class of objects being studied

Probability measure of chance; measures the likelihood of an event occurring

Proportion equality between two ratios

Pythagorean theorem theorem stating that the square of the longest side of a right triangle is equal to the sum of the squares of the remaining two sides

Quadrilateral polygon with four sides; e.g., rectangles and squares

Range largest value minus smallest value in a set of data

Ratio of two quantities first quantity divided by the second quantity

Ray half-line together with one point

Right angle angle formed by lines perpendicular to each other

Right triangle triangle with one right angle

Sample subcollection of a population

Sample space listing of all possible outcomes for a given experiment

Satisfy an equation to make the equation a true statement

Simple closed curve a curve that can be drawn so that it begins and ends at the same point and can be drawn without retracing any of its points

Simple curve figure that can be drawn without retracing any of its points, except that endpoints may coincide

Solution the number that, when substituted for the variable in an equation, makes it a true statement

Solving an equation finding all solutions for a given equation

Square root of x (denoted by $\sqrt{x}$) the number a, so that $a \cdot a = x$

Standard deviation measures the deviation of values from the mean

Term single number or a product or quotient of one or more variables and/or numbers

Triangle polygon with three sides

Trinomial polynomial with three terms

Unlike terms the opposite of like terms

Variable letter used to represent a number

INTRODUCTION

If you consider yourself to be an otherwise competent adult, but a math illiterate...

If you associate mathematics with feelings of anxiety, defeat, and failure...

If you equate doing math with feeling stupid...

If your job involves formulas and calculations, but you're too embarrassed to admit that you don't understand them...

If you really hate math and believe you always will...

You may be one of a vast number of adults suffering from math anxiety.

Those mental blocks you experience while doing math problems may not be a sign that you are mathematically incompetent. Instead, they may be an indication that your attitudes and feelings toward math in general are affecting your ability to concentrate, and to apply what you have already learned.

The frustrations, confusion, and tension often associated with attempting math have finally been recognized as symptoms of a severe and pervading problem, namely *math anxiety*.

Math anxiety is a relatively new topic in the area of self-awareness and self-improvement. However, it is receiving a lot of attention from educators, psychologists, and mathematicians.

Magazine and newspaper articles and even talk shows have devoted major segments to the issue of math anxiety.

Just what is math anxiety?

What causes it, and who suffers from it?

What effect does math anxiety have on self-confidence, daily experiences, career choices, and job advancement?

Finally, how can one relieve math anxiety?

If math has always been a sore spot for you, these are questions you will want to answer.

Math anxiety can be described simply as a fear of figuring, a fear of doing anything mathematical. This description usually brings to mind the stress associated with doing some difficult calculus problem.

Math anxiety is by no means restricted to the typical classroom or test situation. Filling out tax forms, balancing a checkbook, determining interest, buying a new home, doing comparative shopping, understanding your phone bill, setting up a retirement plan, and buying insurance are just some of the day-to-day experiences that often give rise to the tensions and fears characteristic of math anxiety.

For some adults math anxiety may consist of vague feelings of self-doubt concerning their math abilities. These individuals often arrive at correct answers to math problems. However, they lack confidence in their own abilities and attribute their success to luck.

On the other hand, some math-anxious adults panic at the sight of a math question. Percentages and ratios baffle them, and the thought of calculus is terrifying. For these individuals, math anxiety takes the form of great stress, confusion, and a sense of loss of control. People's anxieties prevent them from thinking clearly or even taking the first step toward solving their problems.

Math anxiety can also occur at any of the various stages between these two extremes.

Math anxiety is sometimes referred to as *mathophobia*. Indeed, some of its characteristics are very similar to those of other phobias. For instance, the fear, acute tension, and even nausea experienced by the claustrophobic on a short elevator ride can be very similar to the anxieties and overall helplessness felt by many adults when attempting math-related tasks such as filling out income tax forms.

However, *phobia* may not be an accurate term to describe math anxiety. According to Webster, a phobia is "an exaggerated and usually inexplicable and illogical fear of a particular object or class of objects." This definition would therefore imply that if math anxiety is a phobia, there are no rational reasons or explanations for it. However, I think it's only natural to be anxious about attempting to do problems you don't understand or know you have not been able to succeed with in the past. Perhaps algebra was always a mystery to you, with solutions seeming to appear magically at the bottom of the page in your text. You certainly cannot feel comfortable with materials you do not comprehend. You are not going to enjoy doing math-related work if you have a history of failure with math or math courses. You probably dread it.

Your lack of success with math may be caused by any one of several factors: a poor math instructor at some point, an insufficient number of math courses in high school, unintelligible texts, or misinformation about what math is, what it isn't, and who should do well in math. It appears that there do exist logical and reasonable causes for math anxiety and poor math abilities. However, most adults do not attribute their math difficulties to these reasons. Instead, they blame math failures on their lack of a math mind, the notion that men always do better in mathematics than women do, or that they have poor memories or learning disabilities.

This aspect of math anxiety may be considered characteristic of a phobia. Most math-anxious adults display an unrealistic and sometimes illogical attitude toward their failures with math problems and their subsequent lack of math capabilities. They believe that math is not for them and never will be. This belief is so strong that it prevents them from taking the first steps toward analyzing what their math difficulties really are and how to go about remedying them.

Often, the math-anxious believe that there is no remedy. If they were not able to do math in high school or college, they are certainly not going to be able to learn it now. These individuals really feel that they are too dumb or inadequate to do math. They just don't have a math mind.

Until recently, the idea of a math mind was almost universally accepted by mathematicians and nonmathematicians alike. It was assumed that one is either gifted with talents for doing math or one is not. The mathematically gifted have minds that think clearly and logically, understand new ideas at once, and have an uncanny ability to solve difficult puzzles and problems with both speed and accuracy. They are always the ones playing chess at social gatherings. Traditionally, these gifted individuals have been a small minority of our population, usually made up of men who went on to become our nuclear physicists, chemists, engineers, and of course, mathematicians. For the have-nots, with respect to math minds, it is assumed that one's potential must lie in verbal abilities. Standardized tests seem to reinforce the division between verbal and mathematical talents.

However, even though it is recognized that some have stronger verbal abilities than others, it is assumed that everyone should be able to develop a certain degree of competency in verbal skills. Furthermore, everyone should be capable of applying these skills to daily situations requiring abilities in reading and in writing.

Yet, when it comes to math abilities, you either have them or you don't—or so people once believed. The concept that most basic math skills can be developed by most individuals has developed only recently with the increased research into math anxiety.

Studies and programs in math education have shown that succeeding with math problems is more often determined by one's attitude and feelings toward the subject than by any innate aptitude for math. These studies have also shown that negative math attitudes and the resulting anxieties are not limited to any one group of individuals. College professors, lawyers, social workers, housewives, skilled laborers, and psychologists have admitted that they suffer from math anxiety. Even tax lawyers and economists may feel inadequate in doing certain types of math. For these adults, math deficiencies are even more embarrassing because everyone assumes that they should be mathematically oriented.

Although math anxiety is not limited to any one career group, it does seem to affect a greater number of women. Research by Elizabeth Fennema and Julia Sherman has shown that there is little difference in the ability of males and females to do math. Yet there is an enormous gap in the numbers of men and of women who study mathematics beyond the elementary level. One study by sociologist Lucy Sells found a large discrepancy in the backgrounds of male and female freshmen at Berkeley. Her results showed that 57 percent of the boys had taken four years of high school mathematics while only 8 percent of the girls had such backgrounds. Without those four years of high school math, students at Berkeley could not be admitted to seven of the nine schools and colleges on the campus and were ineligible for twenty-two of the forty-four possible majors at the college of arts and sciences. Thus, 92 percent of the women were limited to traditionally feminine career choices: the humanities, counseling, fine arts, and elementary school teaching.

Math is often regarded as a male subject. *Impersonal, objective, calculating,* and *logical* are terms used to describe both mathematics and the stereotypic image of men. Many people believe that women, on the other hand, are more emotional, are people-oriented, and should not have to bother their heads about numbers. Even though no differences exist in innate abilities, studies do show that women underestimate their own abilities to solve mathematical problems.

Math anxiety has major effects on men and women. It can create a cycle of fear and defeat for individuals with math problems. Anxieties about mathematics can lead to mental blocks and confusion when solving a problem. These feelings of loss of control prevent the individual from making any progress. Failure to arrive at an answer creates even greater anxieties and fears. And so the cycle goes on and on, with anxieties mounting at each step. If you believe that you will not be able to solve the problem before you even start, defeat is certain to follow.

Naturally, the most common result of math anxiety is math avoidance. You stop taking math courses, choose jobs that involve little or no math, get someone else to do your taxes, have a friend check the restaurant bill, or assign math-related problems to co-workers so that you won't have to reveal your inadequacy on the job.

What are the effects of this math avoidance? First of all, math avoidance will prevent you from ever developing better math skills. You will always fear it, and daily tasks involving math or related skills will create more anxieties and will contribute to a low self-image.

In addition to lowering self-confidence, the math anxiety-avoidance cycle has other far-reaching consequences. As technology progresses, math avoidance becomes increasingly difficult. Daily tasks—determining gas mileage, figuring out tips, interpreting utility bills, choosing insurance policies, understanding charts and graphs in newspaper articles—all involve math. It's really just not possible to pass on *all* math-related tasks to someone else. We are constantly in need of math skills. Even a trip to the grocery store, with the various discounts and specials, can become an exercise in mathematical confusion. Personal finances, major purchases, such as buying a car, and choosing a retirement plan are all daily experiences that involve not only math skills but confidence in your ability to compare various collections of data (percentages, deductions, decimals, and so forth) in order to decide on a best purchase.

If math skills are necessary for day-to-day routines, they are becoming even more essential for a wide variety of jobs and for career advancement. Careers previously thought to be people-oriented are relying more and more on math skills. Careers in management, the social sciences, counseling, law, business, and government work are only a few of the fields requiring math skills. A degree in any of these fields requires advanced mathematics. Needed math skills may include reading computer printouts, using ratios and percentages, or understanding statistical data. Higher paying jobs often demand an ability to use such skills. Thus, math avoidance also prevents many adults from career advancement or changes.

Finally, math avoidance also may be keeping you from participating in a whole range of activities. Hobbies such as photography and flying, doing your own electrical work or plumbing, or even solving a puzzling problem can provide many rewards, including both pleasurable and financial ones. However, these activities do require math skills.

Let's suppose that you have math anxieties and you realize that they have a strong influence on your self-confidence and many daily activities. Is there a cure? Fortunately, recent studies have shown that nontraditional math programs can relieve math anxiety. Adults can overcome their fear of math, and they can develop confidence and a sense of control when dealing with math. This confidence will allow them to go on to pursue such subjects as algebra, business math, and even statistics. Programs that have proved successful in coping with math anxiety include counseling, group therapy, remedial courses designed for the adult learner, and clinics and workshops where a nonthreatening approach to math is presented.

This book is an outgrowth of one such workshop, the ongoing Math Without Fear adult education program at Georgetown University, Washington, D.C. At the request of the School for Continuing Education at Georgetown, I designed this program specifically for the math-anxious adult. It was one of the first of its kind in the D.C. area.

Encouraged by the success of other math anxiety clinics, I created a six-week workshop course that offered both math and counseling in a relaxed setting. As an instructor in the math department and a recent Ph.D., I felt I was well qualified to relate to others trying to deal with math anxiety. Graduate math programs involve various stages of frustration, anxiety, and discouragement. The idea that math always comes easily to the mathematician or math major is just another math myth. I often thought that Ph.D. programs in math should be called Math *with* Fear!

The main emphasis of the MWF (Math Without Fear) program was on how to approach math and relate it to everyday situations. Participants were encouraged to develop methods of problem solving that worked best for them.

The program was well received and attracted a wide variety

of adults, including grocery clerks, lawyers, women returning to college, clinical psychologists, college students, real estate agents, and legislative assistants on Capitol Hill. The success of MWF exceeded all of our expectations at Georgetown. Within one year, the number of sections of MWF had to be increased. Also, requests for MWF Parts II and III were continually being received. Mathematicians and counselors in the D.C. area began to ask us about the possibility of helping them create programs of their own.

The success of MWF certainly seems to reaffirm the claim that a large number of adults from a wide variety of backgrounds suffer from math anxiety. Furthermore, the demand for programs such as MWF appears to be growing. This demand also is accompanied by a demand for nontraditional books and resources aimed at the adult learner.

In conducting the MWF program, I found it necessary to develop my own materials. There was no one book that met all of the needs of the class. Books intended to give a thorough understanding of basic math skills usually are too elementary for adults. The adult studying math has years of daily math experiences to refer to. Experiences in paying taxes, using charge accounts, and figuring out a budget all give the adult a great deal of insight in doing math-related problems. Elementary books ignore this added experience.

On the other hand, traditional college and review texts, with their formula-problem-answer format, gave few explanations and completely ignored the effects of anxiety and poor math attitudes. In addition, the few books aimed at understanding math anxiety served as excellent references, but failed to include a sufficient supply of the math problems and skills needed by the adults in the program.

This book is a collection of the materials I developed for the MWF program. It includes the techniques and approaches I found most successful with the program. Advice on understanding and dealing with math anxiety is spread throughout the book. This advice is based on both tips and hints of my own and the ideas and experiences of the hundreds of adults who took the course. The book covers the major math topics most requested by these adults. These topics include a review of basic concepts in arithmetic, algebra, methods for solving equations, working with ratios and percentages, geometry, followed by a basic introduction to statistics and probability.

In the MWF program, I found that after overcoming math anxieties, the skill needed most was the ability to apply math to everyday situations. The word problem seemed to be the key to doing these applications. Word problems also seemed to provide the most anxieties. For many of the adults in the program, solving equations or multiplying fractions could be mastered very quickly. But when it came to taking a verbal problem, or even worse, formulating one of their own and solving it, panic and fear would immediately set in.

The entire course was built around solving everyday problems—problems on the job and in the home, consumer problems, and also classic word problems found in standardized tests.

This book uses the same approach. After discussing math anxieties and attitudes, it begins with verbal problems and how to approach them. Basic math skills are presented as they are needed in the context of the everyday word problems. Adults in the program found that this nontraditional approach had several benefits.

Word problems gave a motivation for introducing basic math skills. Calculating interest and salary raises certainly gave one an incentive for studying percentages. Also, I found that most math-anxious adults feel more comfortable with verbal materials and explanations than with symbols and technical notations. Replacing the traditional drill-equation format with a verbal one seemed to be less threatening. Finally, word problems can be a good starting point in building math confidence for adults coming from a wide variety of math backgrounds. Adults in business, electrical work, social work, and consumer affairs have a common need to be able to take math skills and relate them to daily situations.

I have included word problems representing a wide variety of everyday situations. These problems range from tripling recipes and doing comparative shopping to calculating discounts and understanding statistical data. Step-by-step explanations and solutions have been provided for a large number of these problems. Memorization has been kept to a minimum. In its place I have stressed the importance of developing a positive approach to solving math problems. Realizing that math anxieties are not unusual can be the first step toward building math confidence. Developing a more realistic view of math and mathematicians is the next major step toward building a positive approach to math. This can be done by dispelling math myths such as the notion that one is born with a math mind.

Many math-anxious adults need and should even demand an approach to math that reflects their particular experiences and goals. I feel that these goals were best summed up by one member of the MWF program who said:

> I would like to overcome a lifelong dread of
> math. I need to build a foundation for
> understanding how to approach it. But I need
> an approach different from the traditional
> high school approach and without the laborious
> processes at the elementary level!

If these goals reflect goals of your own . . .

If career advancement means doing more math . . .

If you want to take a math course, but need a good review . . .

If math anxiety forces you to rely on others to handle everyday problems . . .

If you've always feared that math was not for you . . .

Reading this book may be your first step toward building math confidence and enjoying the practical use of mathematics.

CHAPTER 1
SUCCEEDING WITH MATH–ATTITUDE VS. APTITUDE

"Mathematics? Ugh! Always hated it. I just don't have a mathematical mind!"

Mathematics seems to have the distinction of being either loved or hated. There are few who can take it or leave it. Unfortunately, the above quotation seems to be typical of the vast majority of adults. The word *mathematics* is associated with a whole range of negative feelings—feelings of inadequacy, helplessness, lack of control, frustration, tension, and, of course, anxiety.

These feelings often arise when adults recall past math experiences: failing an algebra exam, struggling through a math course, giving up on understanding tax forms, being sent to the board in high school, wading through a series of incomprehensible questions on standardized tests. The list could go on and on.

From past experiences most adults have discovered that math is often dull if not boring, confusing, and obscure, difficult and sometimes impossible to understand. However, these adults don't attribute their unsatisfactory math experiences to poor textbooks, inadequate instructors, or an insufficient number of math courses. For the most part, they blame their own innate lack of skills for their failures with mathematics. They don't have a math mind. They believe that only those gifted from birth can possibly hope to unravel the secrets of mathematics. These gifted individuals, a small and distinguished group usually made up of men, have uncanny abilities to think quickly, reason logically, and do complicated calculations in their heads with amazing accuracy. This is the view that too many adults hold of the mathematician.

The concept of the math mind is not only a math myth, but it is probably one of the major causes of math anxiety for most adults. If you were one of the have-nots when it came to handing out math brains, then it never would be possible to succeed in math. The result of this type of thinking is that you give up before you can even begin to solve a math problem. You expect to fail. Then the first sign of difficulty or confusion in doing such problems seems to reinforce this belief that you are simply not capable of doing them. In other words, you feel that you're too dumb to do math.

Surprisingly, these kinds of feelings are characteristic of individuals who are not only intelligent, but have been extremely successful in other fields. They may even have been able to do well in math-related courses.

I recall talking to a woman who has a Ph.D. in languages, an M.B.A. and a host of awards and scholarly attributes. When I mentioned that I was in math she began to tell me about her experiences with math. She had been able to earn high grades in math classes, but she claimed that she didn't have any real ability to do math. Somehow she had gotten by on memorization and understanding instructors. She then went on to admit that if she had had math ability she could have become a medical doctor, like her sister. It seems that her sister was the mathematically gifted one in the family. This woman went on and on, describing her inadequacy in math and how she felt incompetent with it. It was obvious to me that this woman was extremely intelligent. Both her background and conversation indicated this. However, her problems with math seemed to be dominating our conversation. Frankly, I was interested in learning more about her accomplishments in languages and business. Her conversation impressed me in several ways. First, I was impressed by the fact that this woman really believed she wasn't capable of doing math, despite all her other scholarly successes. Secondly, she allowed her feelings of inadequacy in math to detract from her general self-confidence. Furthermore, this lack of self-confidence had prevented her from making certain career choices.

The concept of the math mind is one of the major myths of mathematics. Success in doing math is due more to self-confidence than to any innate talent or ability. Your belief in your ability to do math will allow you to go past those first stumbling blocks when a problem puzzles you. Perseverance is the key to solving many math problems. Most professional mathematicians will agree that their successes with math have been due to hard work, perseverance, and the belief that if they work long enough the answer will come. In addition, they will probably agree that their careers have involved various periods of frustration, self-doubt, and confusion.

In the MWF program we found that, once anxieties and emotional blocks toward math were removed, almost everyone was able to succeed with math goals. These goals included taking math courses, applying math on the job, and making career changes. Attitudes and feelings toward math have a great deal to do with success or lack of success in doing math.

In addition to inhibiting math abilities, math anxiety leads to even more self-defeating feelings. As one would expect, math anxiety leads to math avoidance. As a result, the mathematically incompetent must rely on the skills (and honesty) of those who are mathematically capable of figuring out their problems. These problems may include determining how much we owe in taxes, what kind of insurance we should carry, what our bank interest is, or which item is the best buy at the store.

The frustration and anxiety caused by our inability to do math is added to by the fact that we are forced to rely on others for various daily tasks. This just adds to the poor image we have of ourselves and increases the self-defeating attitude we have toward math. Lack of self-control and self-confidence seems to grow and grow.

With the feelings of frustration, tension, defeat, and inadequacy associated with doing math, it's no wonder that many adults not only fear, but really dislike math. Some extend these feelings about math to the mathematician. At any social gathering the usual response to the statement that I'm in mathematics is this: "You're in math! Oh, I never could do math! You must be a whiz!" Then the individual moves on to talk with someone else. This reaction has many implications. Note the similarity to the quotation at the beginning of the chapter. If math is thought of as abstract, dull, and difficult to understand, then the mathematician must be also. As soon as new acquaintances discover my background, they hurriedly move on to someone else before I embarrass them by reciting unintelligible equations or quizzing them on their knowledge of percentages. If you think I'm making a case for the social life of mathematicians, you may have a point. But, for many adults, their feelings about mathematicians are intertwined with their feelings for math itself. For the math-anxious, there is a great distance between their self-image and the stereotype of the mathematician. As Dr. Stanley Kogelman and Dr. Joseph Warren, authors of *Mind over Math*, point out, in order to overcome math anxiety and build up confidence it is necessary to have a realistic view of math and mathematicians (see Bibliography).

Many adults are surprised to learn that mathematicians often find math problems puzzling and incomprehensible. The Ph.D. student often views the math dissertation as one very long exercise in frustration. Progress is slow and tedious, and tensions mount. This certainly is a very different picture from the stereotypic image of math majors: no brilliant deductions popping to mind, no dazzling and intricate calculations. Working on a dissertation or any advanced math problem requires slow reading and rereading, attempts and more attempts, and going back over and over again to the same problems. This slow process also is accompanied by numerous anxieties and tensions. As Dr. Kogelman and Dr. Warren agree, just about all mathematicians have difficulty with math at one point or another, and almost all have experienced anxieties in doing problems. Somehow they have developed methods to overcome their anxieties and gain control over math problems. They don't let their anxiety control their ability to concentrate and to progress.

For most adults difficulties with math are not due to learning disabilities. However, the anxiety interferes with concentration and memory and prevents them from doing math successfully and confidently. Attitude, not aptitude, is most likely the source of your math anxiety problem. Thus, relieving math anxiety and building math confidence begins with developing realistic and positive attitudes toward math.

The first session of the MWF program always was devoted to changing negative feelings about math to positive ones, much like going from negative numbers to positive numbers. But we will come back to that later. The rest of this chapter is devoted to the techniques and steps that were most successful in the program.

BUILDING A POSITVE MATH ATTITUDE

Misery Loves Company: Realize That Math Anxieties Are Normal and That Your Situation Is Not Unique

The first meeting of MWF always was informal. The first goal was for everyone to get acquainted. Then members of the workshop shared previous math experiences with each other, myself included. Realizing that others feared math, had trouble doing math, and found it frustrating usually was the first step toward reducing math anxieties and stress. It can be your first step in building a more confident attitude toward math. I've included some of the typical responses of adults in the program who were asked to talk about their feelings and experiences with math:

"I always assumed I would fail at math, and I was always put into classes with people who all seemed to understand except me."

"I freeze when someone gets the solution to a problem before I even understand it!"

"My mind goes blank when I see a math problem." (By far, this was the most popular response!)

"Word problems have always stumped me. Eventually, the sight of them caused panic."

"Fractions, ratios, and percentages—they always seem to baffle me."

"I'm a math illiterate. I don't trust myself beyond adding and subtracting."

"My husband says I'm afraid to try math and he's right! That's why I'm here."

"I freeze when presented with problem solving: Joe is four years older than Moe, who is nine months younger than Smoe. Ugh!"

"Math is a weak point of mine. I get anxiety attacks from it. I don't even know how to use a calculator!"

The adage that misery loves company seems to hold true for the miseries associated with doing math. Somehow, the idea that you're the only one or one of a few with a certain problem makes the problem that much worse. Finding out that your problem is not unusual, that more people suffer from it than not, seems to bring an almost immediate sense of relief. If tax accountants and Ph.D.'s suffer from math anxiety, your own anxiety can't be unfounded. Many adults, feeling unique in having math anxiety, are embarrassed to admit their problems in doing math. Recognizing that almost everyone has a certain degree of math anxiety is reassuring. The misery-loves-company concept is certainly not new. Alcoholics Anonymous and Weight Watchers have used this technique for years. Furthermore, learning that others have been able to overcome their anxiety and succeed with math can give you a glimpse of the possibility of succeeding with math yourself.

Know What Math Is and What It Isn't

Developing a more realistic view of math and what it entails also can be an aid in reducing anxiety. There are various myths associated with doing math.

The first math myth that comes to mind is the one we've mentioned before: the *math mind*. Math involves certain techniques and skills that can be developed by most individuals. This development will require practice and time just like learning to play an instrument, learning to swim, or mastering a foreign language. Many feel that math requires no studying—you either get it or you don't. Obviously, understanding the material is one of the major essential steps in doing math. However, it is not the only one. Practice in doing numerous problems is needed in order to apply the material.

Another myth is that doing math means doing it quickly. Most of us envision mathematicians as scientists solving complicated equations in their heads at lightning speeds. This is hardly the case. Most mathematicians need pencil and paper, often work slowly to catch specific details, and often have only average memories. Because of these characteristics, many mathematicians experience anxiety when asked to solve a problem on the spot or in front of a large group of people. Their anxiety is even greater than that of the average individual since they are expected to be proficient in math abilities. It's assumed that doing calculations mentally is one of these abilities. Furthermore, they're supposed to do these calculations quickly, without hesitation. Granted that some people do grasp some concepts more quickly than others. For most, however, including many mathematicians, certain concepts can be understood only after reading and rereading them several times.

Speed is not the goal in doing math: understanding and interpreting answers are the goals.

Doing calculations quickly in one's head is a technical skill, much like being able to memorize all the numbers on a page of the phone book. Both skills are impressive, but they are of little use in developing and applying skills to everyday or job-related problems.

These are just some of the myths associated with math. We'll refer to other math myths throughout the rest of this book. Dispelling math myths coincides with building a more realistic view of math and a more realistic expectation for your own math abilities.

Be Relaxed and Set Up
a Reasonable Schedule for Math Work

Don't attempt to do math when you're tired, tense, or already anxious from other daily activities. These other tensions will keep you from concentrating and thinking clearly and will heighten anxieties you have about math. In the MWF program we serve refreshments to help put everyone at ease. You might try doing the same or playing some relaxing but not too distracting music. Also, you should work on math when you have a good supply of energy. Math requires concentration, deep concentration at times. If you're already bushed from a hard day at work or home, you'll hardly be able to devote yourself to solving math problems. Finally, find a spot where you won't be disturbed. This may involve separating yourself from other members of the family or co-workers. Math requires being able to think continuously about a problem at least for a short period of time. Distractions will mean that you have to keep going over the same material two or three times, and you may end up losing your train of thought.

I don't want to convey the idea that you need to spend hours at a time doing math problems. Far from it! In fact, it is often preferable to work on math every day for a short period than to work for hours one or two days a week. Doing math only once or twice a week may result in your forgetting some of the material from the past week. Thus, the first half of your study time will be spent reviewing unnecessarily. Also, working for several hours at a time is tiring and naturally leads to reduced performance and boredom. Working a small amount each day may be more beneficial. However, do try to make sure that these work periods are free from distractions. Stopping every ten minutes or so to go to the door, answer the phone, or talk with a friend would prevent anyone from making reasonable progress with a math problem.

Write a Math Autobiography

Many adults find it is enlightening to determine just where their math anxieties and troubles began. By writing a math autobiography you may also be able to pinpoint the source of your math anxieties. Your difficulties in doing math may have been due to a poor text, a course you were never able to take, or a poor instructor at some point in your math background. In a math autobiography you try to write down some of the math experiences, both positive and negative, that stand out in your memory. In organizing your autobiography you may find some of the following unfinished sentences helpful:

I used to enjoy math until_____

_____

_____

_____

I did well in math until_____

_____

One particularly good math teacher I recall was_____
   because_____

_____

One particularly poor math teacher I recall was_____
   because_____

_____

When I think of math, the first words that come to mind are_____

_____

I (always, never) expected to do well in math.

I always assumed only_____(people) could do math.

In schoolwork I usually spent (more, less, about the same amount
   of) time on math as I did on my other subjects.

I would describe my attitude toward math as_____

My mind goes blank when I attempt_____math problems (ratios, percentages and so forth).

I always thought I (would, would never) use math outside of
   school.

   Putting a check beside statements that you agree with below
may also reveal some of the characteristics connected with your
math anxieties:

____I never expected to use math in my job.
____I freeze when I see word problems.
____Math and I just don't agree.
____I always assumed that only men did well in math.
____I usually let my (husband, wife, friend) handle all my affairs
   that need math skills.
____I let my accountant take all responsibility for my taxes.
____I would switch jobs, but advancement means learning more
   math.
____I'm just too slow to do math in my head.

   This autobiography is a variation of the one used by Sheila
Tobias in her successful math clinic at Wesleyan College. I found
it very useful for adults in the MWF program to begin to evaluate
and analyze their own backgrounds.

Do Not Try to Do Too Much Too Soon

The above advice is again a common-sense idea that most individuals apply to other areas of their lives, but forget when attempting math. You can't expect to be able to immediately learn calculus if algebra is still giving you trouble. Reviewing or taking a course that requires material that you have not quite mastered will be frustrating and anxiety-building. Trying to prepare for Law Boards, SAT's, or other standardized tests without a sufficient review of necessary math skills also will be frustrating.

If you are reintroducing yourself to math, begin with problems that you know you can succeed with, even if this means going back to addition and multiplication. Beginning with problems you know how to solve will give you the positive results needed for building a good math attitude. Each successful math experience helps to reduce anxiety and build math confidence.

Realize that math takes time. Doing math successfully does not necessarily mean doing it quickly. It involves understanding, reasoning, and even intuition.

This process entails going from one step to another and sometimes can go quite slowly. This is natural! It takes time to read a problem, more time to understand it and then to solve it.

Mathematics is one area where speed reading is not helpful. In fact, mathematicians are often very slow readers in general because they are used to reading math texts and journals, where each word or symbol is significant. Understanding such material requires at least two or three readings. So don't worry if you feel that you're working at a slow pace. As you progress and do more and more problems, you'll pick up speed. However, your major goal should be to understand. It doesn't help to arrive at answers in seconds if they are incorrect or you are not quite convinced that your method is a correct one.

Enlist the Help of Others

Don't be afraid to ask others for help and advice with math problems. Perhaps your spouse, a good friend, or even your children may be happy to work with you.

Do be wary of having someone else do problems for you.

You learn math by doing math.

In addition, having others do problems for you may be confusing. They may use methods you are not familiar with and may have difficulty explaining their procedures. What they can do is listen and ask questions while you do problems for them.

You learn more math by trying to explain it to someone else than by sitting in any number of math classes. Often you think you understand a concept completely until you find loopholes while trying to explain it to someone else.

You also may want to work with another adult who has math anxiety. Meeting with this person once or twice a week to share ideas and problems (and feelings) can be very beneficial. No two people approach a problem in exactly the same way. You can both pick up insights and ideas from each other. "Two heads are better than one" is a saying that often applies particularly well to doing math. The idea that a math problem can be solved in one way and one way only is another math myth. The fact that many adults believe there is only one way to do math is unfortunate. If anything, solving math problems in new or different ways should be encouraged. Mimicking another's method will only be an exercise in memorization. Furthermore, relying on this technique will lead to failure when a problem is slightly changed or when you want to apply the method to a more difficult problem.

Each adult needs to find a method that works well for him or her. This method should be one that you can understand thoroughly and one you can apply to various problems. Your method may involve making outlines, drawing pictures, or even counting on your fingers.

Different methods work better for different people. This concept has been very popular in other areas. Just look at the enormous variety of diets available! For some people, skipping breakfast and lunch is the ideal method of losing weight. Others rely on five or six small snacks each day to keep their waistlines trim. Still others stick to the three square meals to control eating urges.

Succeeding in any endeavor means developing the method that works best for you. Do enlist the help of others to give hints, tips, and explanations that may be helpful in building your own method. Just be careful to avoid situations in which someone else does all the explaining. Self-reliance and self-confidence are your major goals in succeeding with math.

The ability to do math does not mean you have to be born with a mathematical mind or an innate aptitude for math.

Success in math depends heavily on your own attitude toward math and on confidence in your ability to solve math problems. Math anxieties can be reduced by developing a realistic view of math and mathematicians. This means dispelling many popular math myths.

Finally, becoming competent in math entails building your own method of approaching math problems while developing a positive attitude. Both of these goals make up the major themes of the following chapters.

CHAPTER 2
WORD PROBLEMS–
WHY ALL THE FUSS?

One of the most enjoyable aspects of the MWF program was the variety of adults attracted to it. We had men, women, housewives, executives, lawyers; just about every profession was represented. They brought with them their different backgrounds and viewpoints that made our discussions interesting.

However, this same aspect presented us with a problem. These diverse backgrounds also represented a wide range of math experience and ability. For some, arithmetic was still a major obstacle. Others had taken courses in statistics and calculus. However, they all lacked self-confidence in math and experienced anxiety doing it.

I felt that there should be at least one math problem or topic that was troublesome for the majority of these adults. I hoped this topic would not necessarily involve specialized information from specific courses. I didn't have to look far for an answer. There was one area of math universally feared by all in the program: *word problems*. Just mentioning this topic aroused anxiety and increased tension among those in the program.

Almost everyone, at some time or other, has had difficulty with word problems. Many in the program even had their own pet word problems that they disliked the most. For one woman, it was "If John is twice as old as Harry's younger brother who is three years younger than Joe . . ." One man dreaded this problem: "Find the final cost if the item is discounted 15 percent off the sales price, which already was 25 percent off the retail price."

Of all math problems, word problems appear to be the most confusing and the greatest anxiety producers. Many math-anxious adults have found that they were able to master different techniques in math: solving equations, adding fractions, calculating percentages, and maybe even doing square roots. But when it came time to apply these same techniques to word problems, they just didn't know where to begin or even how to approach such problems.

This situation is really unfortunate since doing math and using math means solving word problems. Rarely does an everyday problem or job-related problem occur where all the equations are listed and you are asked to find x! Instead you are presented with a collection of facts, for example your present salary with an estimated raise for next year together with statistics on the estimated cost-of-living increase for next year. Your problem is to determine if your estimated salary will keep up with the rising inflation. You know math is involved here, and probably you have guessed that it will involve percentages. However, you have not been given any equations or formulas; it is up to you to set up the problem.

This process of determining the corresponding math needed can be puzzling and frustrating if you don't know where to begin. Why is it that so many of us have this difficulty in applying math to concrete situations?

I believe that one reason for the lack of confidence and skills to do word problems is that traditional math instruction did not teach you how to approach verbal or applied problems. Instruction usually emphasized math skills and calculations. If you were able to master these skills, it always was assumed that you could make the transition from the verbal problem to the mathematical equations and formulas. Few explanations were given of how to make this transition:

A bus travels round trip from Wilkes-Barre, Pennsylvania, to New York City and back again in 300 minutes. Its average speed going to New York was 45 mph. However, on the trip back to Wilkes-Barre, because of less traffic, the bus averaged 55 mph. How far is Wilkes-Barre from New York?

$$\frac{x}{45} + \frac{x}{55} = 5$$

Not only was little advice given on how to do word problems, but often instructors chose this topic to rush through.

A possible reason for the once-over-lightly treatment often given word problems may have been the very fact that students found them so difficult. Classes could be held up for hours getting bogged down by each individual word problem. Sometimes teachers felt, and many still do, that no matter how much time is spent, some students will master the problems and most will not. So why waste time? This attitude is especially prevalent among teachers who have not been overly successful with word problems themselves. They may be able to arrive at the correct answers, but are not quite confident enough in their own strategies to be able to explain them to others. Or perhaps they understand their own methods but have major difficulties in trying to explain them.

The traditional classroom presentation (or lack of presentation) of word problems brings us back again to the *math mind* myth. If one person was mathematically gifted and another was not, then one was either capable of solving word problems or one was not. For those with math minds the transition from words to mathematical equations was an obvious one. The concept that there are various skills and procedures involved in this translation and, furthermore, that these skills could be developed and refined by the average student, was and is foreign to most instructors.

Whatever the cause, it has been unfortunate that the most useful topic in math instruction, word problems, has traditionally received the least amount of attention. Happily, recent texts on the elementary and high school levels are now devoting larger sections to solving word problems. There are methods and techniques characteristic of solving word problems, and instructors and new books are trying to present some of these techniques.

I certainly don't want to give the impression that there is only one *correct method* of solving word problems! As mentioned earlier, each person will approach math problems differently, depending on background, schooling and experience. For instance, someone interested in horse racing may be very capable of working with odds, while another, who has a hobby of carpentry, may be much more familiar with fractions. The concepts and approaches that work well for one adult probably will be very different from those that work well for someone else. Certainly, no two artists would paint a landscape the same way. In fact, if you compare an impressionist's version with a modern artist's or maybe a cubist's, you may not even realize that the paintings are of the same landscape.

However, artists will probably agree that even though interpretations vary from artist to artist, there are basic skills in drawing, form, and coloring that all beginning artists need to master before developing their own styles. Otherwise, there would be no need for art classes.

These same concepts can be applied to solving word problems. There are basic ideas and procedures needed to build your own method of problem solving.

Solving word problems involves much more than finding solutions to equations or doing calculations. It requires reading for understanding, determining what questions must be answered before arriving at the final solution, translating verbal relationships into mathematical ones (i.e., finally arriving at those magical equations!), obtaining an answer or answers, and interpreting your numerical solutions in terms of the verbal statements of the original problem.

For the majority of adults who perceive doing math as adding fractions, multiplying polynomials, or calculating square roots, it is no surprise that solving word problems can be difficult and seemingly unrelated to the math they have learned. With this limited awareness of all the steps involved, naturally panic and anxiety will arise in trying to do word problems.

This chapter, like the initial meetings of MWF, has been devoted to the word problem. We will be spending time on how to approach it, how to begin to strive toward a solution, and how to keep going when you're stuck. In addition, helpful hints found to be successful in the MWF program will be interspersed throughout. Again, you may find some of these tips more helpful than others.

Remember that the key to doing math is the ability to apply the math you know to everyday situations. This will mean developing your *own* method, one that makes sense to you and one you can succeed with.

GETTING STARTED

Beginning with word problems not only provided a common starting point for the MWF program, but those in the program found this approach to be less intimidating than plunging head on into equations, formulas, and calculations.

I think that a major reason for this response is that most people who feel inadequate mathematically feel comfortable and often excel in verbal areas. In fact, most of us believe that we are either verbally oriented or mathematically oriented. Again, we have the idea of the math mind coming into play. That one could be both verbally and mathematically capable is an idea usually reserved for those in the genius category. Our goal of course is to provide math-anxious individuals with a certain degree of capability and confidence in dealing with math. In other words, we want to build a positive math attitude. Thus, approaching math on a level that most people are comfortable with seems to make sense. New concepts such as equation solving, percentages, ratios, and statistics then can be introduced in the context of word problems. Although this is certainly a nontraditional approach, I believe it's a reasonable and beneficial one.

In MWF we began our strategy for attacking word problems by establishing a basic math vocabulary. After all, isn't math really just another language, like French, German, or Spanish? Math even is referred to as the language of science. I maintain that math is quickly becoming the language of business, government, social sciences, and consumer affairs. So why not study math as you would study French or any other language? One doesn't begin a language by immediately conjugating verbs, memorizing tenses, and translating novels! One usually starts with basic words and expressions such as the names of colors, months, and days, relating the new words to daily experiences. Then one can begin to build simple sentences. For example, in French after learning *mon* for *my*, *chien* for *dog*, *est* for *is*, and *noir* for *black*, one can make the following statement: *Mon chien est noir.* (My dog is black).

We used the same approach in MWF. After determining that *sum* was represented by a plus sign (+), and *equals* was translated into an equal sign (=), "The sum of 2 and 3 is 5" was found to correspond to this expression:

$$2 + 3 = 5$$

Introductory language courses often have titles such as Conversational French or Conversational Spanish. The title is intended to convey the idea that the course will make you feel at ease with the new language while helping you to develop steps toward becoming somewhat fluent in the language. A major goal of these courses is to help the students develop the ability to translate simple expressions and later move on to forming expressions and ideas of their own in the new language. Our goal in MWF is the same, and thus I've entitled the next section Conversational Math.

CONVERSATIONAL MATH

We begin this section by introducing the math symbols associated with various words or verbal expressions. We will be using these symbols in doing translations from word statements into mathematical expressions. Just as the French word *glacé* can correspond to several English expressions, including glass, mirror, chill, and ice cream, many math symbols can be associated with more than one English word or expression. Some of the most common associations corresponding to math symbols are listed in Table I on page 36. The table gives verbal expressions that are commonly indicated by the various mathematical symbols.[1]

You might want to note that this symbol ($\cdot$) has been used to indicate multiplication. Many of us are more familiar with this symbol ($\times$) for multiplication. In doing word problems, the letter X is often used to designate some unknown quantity still to be determined. Therefore, using this same symbol to represent multiplication could be a source of confusion. Thus, the product of X and Y is denoted by $X \cdot Y$ rather than $X \times Y$. This also is a good time to mention another common notation used to denote multiplication: juxtaposition. Instead of writing $X \cdot Y$, we often use this shorthand notation: XY. This abbreviation is particularly useful in more complicated mathematical expressions such as this one:

$$3ab - 5bc + 4ac$$
in place of
$$3 \cdot a \cdot b - 5 \cdot b \cdot c + 4 \cdot a \cdot c$$

TRANSLATION TABLE

+	−	·	÷	=	x, y, n
SUM	SUB-TRACT	TIMES	QUOTIENT	IS	UN-KNOWN
PLUS	SUB-TRACTED FROM	PRODUCT OF	RATIO	BECOMES	
ADDED TO	LESS OR LESS THAN	MULTI-PLIED BY	DIVIDED BY	IS THE SAME AS	
IN-CREASED BY	DIFFER-ENCE	TWICE OR DOUBLE, ETC.	ONE-HALF, ONE-THIRD, ETC.	IS EQUAL TO	
GAIN	SMALLER THAN	FOR		WAS	
EXPAND	REMAIN-DER			IS RE-PLACED BY	
MORE THAN	DE-CREASED BY			WILL BE	
GROW	DROP			ARE	
AUGMENT	LOWER				

This notation is more concise and eliminates possible confusion with decimal points.

Using these expressions and knowing how to interpret them will be topics covered a little later in the book.

TRANSLATING SHORT VERBAL PHRASES

Our next step in the MWF program was to practice translating short verbal phrases into mathematical ones. We usually began with a phrase similar to this one:

a number decreased by 4

When translating from English into math your first step should be to determine if there are any unknown quantities or elements involved. In this example we have one unknown element, namely the number. We would like to represent this item by a symbol. I chose the letter n. There is no requirement to use particular letters to represent unknown quantities. In fact, most books probably would use X or Y to denote the number. I've always felt that if you use a letter closely associated with the word it represents, the notation then helps reinforce the purpose of your problem. Sometimes we get so involved with the symbols that we forget what they stand for and lose sight of the corresponding answers to the verbal problem. Thus, n for number, seemed to make more sense to me. Many of the adults in the program liked this method also.

Since there are no other unknown elements in the problem, we will begin to make a translation using Table I. We note that the phrase decreased by corresponds to this math symbol $(-)$. The symbol 4 is already a mathematical one and needs no translation. Thus,

a number decreased by 4

becomes

$$n - 4$$

(You might be surprised to learn that we are already using algebra! Looks easy enough, doesn't it?)

You probably have found this first example to be fairly simple. Good! Even though this was a simple case, it's precisely this type of translation that turns out to be the key to solving word problems. Let's look at another example:

The sum of three times a number and four times the same number.

In the above phrase we again have only one unknown element, the number. We'll use the same notation as before, n to represent it. Next we note that we are talking about a sum. A sum of what? In this example we are summing, or adding, two items. The first item is three times a number. This expression can be represented by $3 \cdot n$ or, using the abbreviated notation, $3n$. What is the second item being added to this first quantity? The answer to this question is "four times the same number." This second quantity can be denoted by $4 \cdot n$ or simply $4n$. Finally, we note that the sum of these two items can be represented by the math symbol for addition $(+)$. (Again we are referring to Table I for these translations.) Thus, the sum of our two items can be denoted this way:

$$3n \ + \ 4n$$

Notice that in this example we are not able to take each word and go immediately to the mathematical translation. However, by breaking the verbal phrase into parts, translating each part, and then representing that final sum, we are able to get our final mathematical representation. This procedure actually involves rewording the problem into your words. You must ask yourself just what is in the sum, how you will go about representing these items, and you must note that they are to be added together. This method of rewording and rethinking the problem was found to be especially useful in MWF.

Our third example is going to be just a little more complicated in that it involves more than one unknown element:

Twice the sum of two numbers, increased by 5

After examining the above verbal expression we note that there are two unknown items, the two numbers. If we let n represent one number, we might choose m to denote the other. We might also note that a comma divides the phrase into two parts. We'll want to keep these parts separate in the math translation also. To do this, many in the MWF program found it helpful to put parentheses around the expressions involved:

(Twice the sum of two numbers), (increased by 5).

Now we can work with each part separately, using Table I. "Twice the sum of two numbers" implies that we are taking a sum and then multiplying by 2. What sum? The sum is $n + m$, which again you will want to enclose in parentheses: $(n + m)$. Twice this sum is

$$2 (n + m)$$

This resulting number is to be increased by 5, which corresponds to this mathematical symbol: $+5$. We have translated twice the sum of two numbers, increased by 5 into $2 (n + m) + 5$.

Note that without the parentheses, we would have obtained the following representation:

$$2n + m + 5$$

This expression would have corresponded to the English phrase "Twice one number, plus another number, increased by 5." We see that commas do make a difference. Mathematicians have agreed upon a universal order for doing math operations (adding, subtracting, multiplying, and dividing) in order to eliminate as much confusion as possible. We will talk more about this in the chapter on algebra.

Our next phrase could be one applied to an everyday problem: Frank sells antique model train sets. His weekly salary is $180, plus a $60 commission for each train set that he sells. Expressions such as this one are usually previews to problems such as this: If Frank made $750 last week, how many trains did he sell? Answering this question would first involve translating the original phrase into a mathematical one to represent his weekly salary.

When we reread the expression we might first note that his salary comes from two sources. He gets a flat rate of $180. Then, as his commission, he receives $60 for each set he sells. The unknown element in this expression is the number of sets he sold. I've chosen the letter t to denote this number. From experience you've had in making purchases, you probably would realize that the amount made in commissions would be obtained by multiplying $60 times *t*. In addition, we can also note that the word for corresponds to multiplication. (Refer to Table I.)

Thus, "His weekly salary is $180, plus a $60 commission for each set"

becomes   "His weekly salary is 180 + (60t)."

It is to be understood that we are working with dollars. Doing translations such as the one above can involve both rewording the given expression in your own words and doing the actual translation into math symbols. You may want to practice with some of the phrases listed below. Possible translations are given below:

(1) The sum of two times a number and six times the same number
(2) Three times a number, decreased by five times that same number, plus 6
(3) The quotient of three times *b* and 5, decreased by the sum of 5 and *b*
(4) Twenty decreased by the sum of two numbers
(5) Gloria and Harry both had to work last Saturday. Harry worked two hours longer than Gloria. Represent the number of hours Harry worked in terms of the number of hours worked by Gloria.

In the next chapter we introduce translating complete verbal statements. These statements will correspond to math equations, and therefore we will introduce basic equation solving as one of our first math skills.

Answers: (1) $2n + 6n$ (2) $3n - 5n + 6$ (3) $3b/5 - (5 + b)$ (4) $20 - (n + m)$ (5) If we let G stand for the number of hours worked by Gloria, then the number of hours worked by Harry is G + 2.

REFERENCES

1. Selby, Peter H. *Practical Algebra.* New York: John Wiley & Sons, 1974.

CHAPTER 3
KEEPING EVERYTHING IN BALANCE

Keeping everything in balance! The word *balance* can be used to apply to a variety of situations. In this chapter we are going to aim for a balance in two directions. The first will be directed toward building a positive math attitude. We cannot eliminate all anxieties associated with math. There is no guarantee that at some point in the future you will not meet up with some difficult problem that will give you some anxious moments. Having a positive math attitude does not mean that once you have reached a certain point *all* problems are easy. What a positive math attitude will do for you is the following: it will allow you to maintain a balance between natural fears associated with a hard problem and the self-confidence you have developed in approaching math. You may not arrive at the exact answer, or even a relatively correct one. However, math without fear means that you will not allow anxieties to mount and mount till you reach the point at which you can no longer work on the problem effectively. Even mathematicians often find problems confusing, frustrating, and *discouraging*. But they do not allow this discouragement and the associated anxieties of not seeing an immediate answer overwhelm them. Likewise, one of your major goals should be to control your emotional reactions so that they do not prevent you from solving math problems. As you progress through *Math without Fear* you will begin to see a decrease in the amount of anxiety you experience while doing problems. Associated with this decrease in anxiety you will also begin to notice that doing math is not quite as unpleasant as you have always thought. You will even begin to enjoy it. I feel that the only reason so many people dislike, and even hate, math, is that they have never understood it, that they have always done poorly, and that their lack of math ability has prevented them from doing various things. With this kind of history, anyone would dislike a subject. However, these same adults, as they begin to understand what is actually going on and as they begin to succeed with basic problems, become more and more confident.

In Chapter 2 we saw that one way to feel more at ease with math is to become familiar with its language. Many adults experience fears and anxieties with math because they are confused by its various terms, expressions, and symbols. Instead of lacking a math mind, they may simply be lacking a *math vocabulary*.

We began our study of the language of math by acquainting ourselves with some of the symbols and terms used in math, together with their various meanings. Keeping in mind that our major goal in Math Without Fear is to apply math to everyday problems, we saw that this application would require translating verbal problems, or word problems, into mathematical ones. We continued our development of the language of math by translating short phrases and expressions into math phrases or expressions.

In this chapter we are going to go one step further by translating complete verbal statements, or sentences, into mathematical sentences. These new math sentences turn out to be equations, and thus we introduce our first basic math tool: solving equations. This is the second direction of "keeping everything in balance." As we will see later in this chapter, equations are really a balance between two mathematical expressions, each on one side of the equal sign. For example, $2X + 1 = X + 3$ is an equation. Solving equations will involve finding values for X that make the statement, or math sentence, true. In other words, the number we finally substitute for X must maintain the balance between the two sides. For example, if we put 1 in for X we get this:

$(2 \cdot 1) + 1$, or $2 + 1$, which is 3, on the left side of the equation and $(1 + 3)$, or 4, on the right side. For $X = 1$ we do not have an equality.

However, if we substitute 2 for X, we get:
$(2 \cdot 2) + 1$ or $(4 + 1)$, or 5, on the left side of the equation and $(2 + 3)$, or 5, on the right side of the equation also. Thus, $X = 2$ satisfies the equation.

The remainder of the chapter will be devoted to finding a method to show that 2 does make the equation a true statement while other values of X do not. Without saying so directly, we have actually begun working with algebra. Algebra is merely a generalization of arithmetic. In algebra, letters are used to represent unknown numbers or quantities. This use of letters to denote unknown elements in algebra allows us to translate word statements into mathematical sentences.

TRANSLATING COMPLETE VERBAL STATEMENTS

We begin this section by looking at a verbal sentence involving numbers.

Example 1: Seven times a number minus three times the same number is equal to 16.

In this example I chose the letter n to denote the unknown in this sentence, the number. You may recall from the previous chapter that many in MWF found it helpful to use letters associated with the item they are representing, rather than arbitrarily using x or y. "Seven times a number" corresponds to the product of 7 and n, or $7n$. (You may also want to recall that we often abbreviate $7 \cdot n$ as $7n$.)

The word minus corresponds to the mathematical symbol called a minus sign $(-)$. Three times the same number will translate into $3n$. Is equal to corresponds to the equal sign $(=)$. And, finally, 16 needs no translation. We can write the final translation of "seven times a number minus three times the same number is equal to 16" this way:

$$7n - 3n = 16$$

An algebraic statement or sentence such as this is called an *equation*. It represents two quantities that are equal to each other. In this case the first quantity is $7n + 3n$ and the second quantity is 16. Finding the number, which when substituted for n, gives you a true statement, is referred to as *solving the equation*. We have introduced some new words into our math vocabulary. At this point, it probably is worthwhile to stop and introduce a few other definitions used in algebra. (You may want to note that we have *italicized* new math words and symbols.) A letter used to represent a number is sometimes referred to as a *variable*. It also can be referred to as a *literal number*, where *literal*, in algebra, means letter. Thus, *n* is a variable or literal number in our example. A *term* is either a single number, the product, or quotient of one or more letters (variables) and/or numbers. For example, 16, 7*n*, and 3*n* all would be *terms*. An expression in algebra will contain one or more terms connected by plus or minus signs. Thus, 16 by itself also can be considered an algebraic expression. In addition 7*n* + 3*n* is also an expression. We see that an equation actually will be two algebraic expressions set equal to each other. As we have stated earlier, many word problems translate into one or more equations which then need to be solved to answer the problems. Before dealing directly with techniques for solving equations let us continue our discussion of translating complete verbal sentences into mathematical sentences or equations.

Example 2:   A deluxe microwave oven sells for $948. The cost of this appliance is twice the cost of an ordinary stove.

In this example our unknown quantity is the cost of the stove, which we could denote by the letter *s*. The cost of the microwave oven is $948. This cost is "twice the cost of the stove," which corresponds to 2*s*. Since is is associated with the math symbol known as the equal sign (=), we can do a complete translation: The cost of the microwave oven is twice the cost of the stove. This becomes the equation below:

$$\$948 \ = \ 2s$$

Some verbal statements may imply relationships rather than state them explicitly. You may have to rely on previous experiences as a consumer, employee, or student to discover these relationships. For instance, let us look at the next example involving salaries.

Example 3: A man worked for 6 hours and earned $28.20. What was his hourly wage?

We are not going to solve this problem at this point. However, we will attempt to translate the implied verbal relationship into a mathematical expression. What is the unknown in this example? It's the man's hourly wage, which we can denote by w. Now what is the relationship between w and the $28.20?

Almost everyone has worked at some job that paid by the hour. Even the youngest baby-sitter probably could tell us that your total salary is the number of hours you worked times your hourly wage. Thus, $28.20 is 6 times w. Translating we get $28.20 = 6w$. Do not be afraid to write out in words relationships that are explicitly stated, such as our first example. Any extra notes you make will not only help to emphasize major elements of the problem, but will also give you a greater sense of control. Approaching the problem with pencil and paper certainly is a more organized approach than trying to solve the problem in your head.

With an ever growing number of adults becoming calorie conscious, the following example seems to be a timely one.

Example 4: After gaining 12 pounds, George weighed 185 pounds. What was his previous weight?

Again we are just going to set up the mathematical expression representing the above verbal relationship. Once more, the relationship between George's previous weight and his present weight is not explicitly stated. However, from experience in following your own weight, you would realize that 185 pounds is determined by taking George's previous weight and adding 12 pounds to it. Our implied relationship is "185 pounds is George's previous weight plus 12 pounds." Now we can make the translation into math, again by using the letter w (this time to denote weight). We have the following equation:

$$185 = w + 12$$

In our next example it will be necessary to use a fact that you use often in day-to-day experience. This fact is that the whole of any quantity is the sum of its parts. Let us use this idea in the verbal situation below.

Example 5: A 16-inch stick is to be broken into two parts to support two tomato plants. One of the plants is about 2 inches taller than the other, and therefore the stick supporting this plant should be 2 inches longer than the other one. How long should each piece be?

In this example our unknowns are the lengths of the two pieces. We might denote the length of the first piece by L. The second piece is 2 inches longer than the first piece and can be represented by $L + 2$. What is the relationship between these two lengths? We know that their sum will give us the whole length of the original stick, or 16 inches. The length of the first piece plus the length of the second piece is equal to 16 inches. We could represent this situation in the sketch below. We will see that diagrams such as these can be especially useful in solving the word problems in Chapter 4.

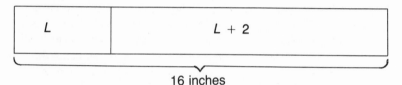

16 inches

Having discovered the implied verbal relationship in our problem, we can state the math translation as follows:

$$L + (L + 2) = 16$$

The rest of this chapter will be concerned with solving equations such as this one.

SOLVING SIMPLE EQUATIONS

You probably have noticed that we have broken up our presentation of solving word problems into several short segments: building a math vocabulary, translating short verbal phrases, translating complete sentences, solving equations, and culminating in a plan of action to tackle word problems in Chapter 4. We have taken a rather large project and attacked it by breaking it up into smaller parts, presenting each part and then progressing to the next— one step at a time. You will want to use this same idea when working on math. Trying to tackle too much at one time can be frustrating and tiring. You will probably have little enthusiasm for a project that seems too large and too hard to complete. Break up your work into smaller parts: concentrate on one part at a time. Then when you have successfully completed that section, you can progress to the next step at a later time. For example, if you want to solve twenty problems, work on five at one sitting. Particularly if the problems seem a bit difficult, attempting five problems will not be nearly as intimidating as trying to do twenty. Furthermore, you are more likely to complete the five problems. If you think you have twenty to do and that they may take hours to complete, you may stop at the second or third problem that gives you trouble. You could end by dropping the project altogether. If you are doing review problems, trying to master a certain area such as fractions or any math project, break up the project into smaller parts. Work on the project a step at a time. Most in MWF found this approach to be less intimidating and more successful.

Before continuing with tools for solving equations, listed here are verbal sentences you can translate into mathematical equations. Possible translations are given below. Recall that the choice of letters will vary from person to person:

(1) Five times a number decreased by 2 is 18.
(2) Six times what number is 24?
(3) The sum of what number and twice the same number is 27?
(4) Three times some number is five less than the same number divided by 2.
(5) Judy spent $67 on her new food processor. This is $13 more than Tom spent. How much did Tom spend?
(6) Dan's weekly pay is $345. What is his net pay if $115 is deducted for taxes?
(7) Bob is five times as old as Jimmy. If the total of their ages is 12, how old is Jimmy?

Possible translations:
(1) $5n - 2 = 18$
(2) $6n = 24$
(3) $n + 2n = 27$
(4) $3n = (n/2) - 5$
(5) Let T stand for the money spent by Tom. Then the money spent by Judy is equal to $T + 13$. Thus, $67 = T + 13$.
(6) $P = 345 - 115$

(7) Let B equal Bob's age and J equal Jimmy's age. Then B equals $5J$, and the total of their ages, 12, is equal to $5J + J$, or $12 = 5J + J$.

Most word problems translate into one or more equations. Thus, solving equations turns out to be one of the most vital skills we need to develop in order to work with practical, everyday problems. Our translation techniques from Chapters 2 and 3, together with some practice in solving simple equations, will be essential for developing your plan of action for approaching word problems.

This plan of action will be your own organized approach to the word problem. In Chapter 4 there will be a five-step guide to help you get started with a problem and, just as important, to keep you going!

Again, in Chapter 4 we will restrict ourselves to some basic types of word problems. One of the major themes of the Math Without Fear program was to progress from the simple problems and techniques to the more complicated ones.

As we mentioned earlier, one of the major obstacles confronting math-anxious adults is that they try to do too much too soon. Reintroducing yourself to math by beginning with difficult problems or examples can have as many negative effects as trying to do too many problems at one time. First of all, it is likely that you will have little success.

All of mathematics can be viewed as a series of building blocks, with each new topic being defined in terms of previous material. Each new step builds on previous ones. Without a solid foundation of the basics, new topics or problems will be confusing, methods to solve them will be vague, and you will have little confidence in your ability to tackle them.

Attempting very difficult problems in your reintroduction to math tools can prevent you from succeeding. Second, tackling difficult problems in the beginning can give you a poor outlook toward your project of becoming self-reliant and self-confident in math. You certainly will not be able to enjoy it. Any task that appears difficult will most likely be one that you will dislike and, naturally, put off doing or aviod altogether. However, attempting simple examples and succeeding with them will give you a sense of accomplishment, even if you consider the examples to be easy. With each successfully completed math example, or problem, you will gain confidence, self-satisfaction, and you will be ready to progress to the next step.

Many in the Math Without Fear program surprised themselves when they were able to solve various math problems. In the beginning of the program one common reaction was "It seems so simple; I must be wrong." Math does not have to be difficult. In fact it is only difficult when you do not understand it. However, many math-anxious adults feel that if they get an answer easily, it must be wrong.

As you progress through *Math Without Fear*, your success with each topic will give you more confidence. And, as the adults in MWF found out, answers to problems that seem easy are not necessarily wrong, but reflect your growing ability to approach and solve practical problems. In this chapter we begin to lay our foundation for approaching everyday word problems by beginning with basic equations and how to solve them. Most adults have been faced with equations at one time or another. This section will be your reintroduction to equations. Instead of viewing equation solving as a purely technical skill involving memorizing formulas and calculations, we will use a more intuitive approach. We will connect the idea of two things being equal, equivalent, or the same in an everyday sense, to the corresponding expression or concept in math.

As we saw in our translation techniques in the beginning of the chapter, equations really are mathematical sentences. We translated verbal sentences into mathematical ones. These mathematical sentences are statements that two quantities or expressions are equal. Furthermore, we will see that these equations can be simplified to determine the unknown quantity. This translation and simplifying procedure will allow us to solve a vast collection of problems. This is the purpose of algebra. We will talk more about algebra in Chapter 7. Now, we will be concerned with using algebra for solving basic equations. So far, we have learned that equations are simply mathematical statements saying that two expressions are equal. These expressions probably will involve one or more elements. A typical equation is given below:

$$x - 5 = 3$$

Continuing our development of a math vocabulary, we define a *solution* and what we mean by *solving an equation*. A *solution* to our equation above is the value that we can replace for *x* that will maintain the equality. For example, if we let *x* have the value 8, then $8 - 5 = 3$ is an equality, or true statement. *Solving an equation* is simply the process of finding all solutions, that is, all values of the unknown that make the equation true. Sometimes you will see this same idea expressed as the process of finding the value of the unknown that *satisfies the equation*. In addition, this solution for *x* is sometimes referred to as the *root* of the equation. As we mentioned earlier, it is important to familiarize yourself with the various equivalent expressions and terms involved in math.

How do you go about solving an equation? One way would be by trial and error, substituting all possible answers in the equation until you find the numbers that make it true. Obviously, this method would be cumbersome and ineffective. A more sensible approach is to find another equivalent equation whose solution can be seen at a glance. *Equivalent equations* will be equations whose solutions are the same. For example, $2x = x + 1$ and $x + 2 = 3$ are equivalent equations. In both cases, the value of *x* that makes the equation true is 1. That is, $2(1) = 1 + 1$, and $1 + 2 = 3$. Each equation has this solution: $x = 1$.

An equation will be solved if we can finally write it in this form: $x = \square$, where $\square$ is some real number.

Our next step is to find a method for getting equivalent equations until we finally reach the step where $x = \square$. In the Math Without Fear program we found it especially helpful to think of an equation as a set of scales, pictured below.

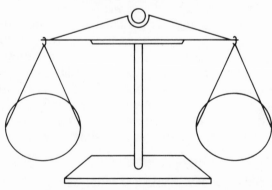

In order for the scales to remain balanced, equal weights must be placed on each side. In the same way, the left member of an equation and the right member must represent the same number. To solve equations we will treat every equation as a balance of the two sides. For example, we could express $X - 5 = 3$ by the diagram of a balance below:

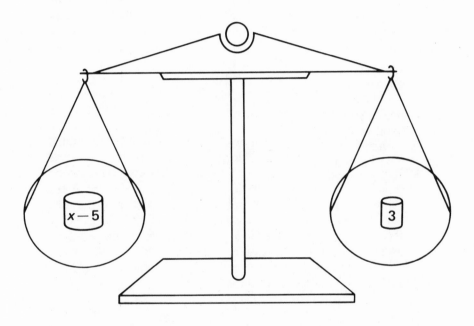

If we want to derive an equivalent equation, any changes we make must not disturb the balance. Any amount or number we add to one side of the equation must also be added to the other side. Continuing with our example, we can get an equivalent equation by adding 5 to both sides. This will help us get to the stage where $x = \square$.

$$x - 5 + 5 = 3 + 5$$

In our diagram, this means that we have placed another weight representing the number 5 on each side of the balance:

Now we have this balance:

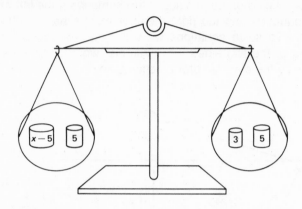

   We know that $-5 + 5$ is zero. (You may want to refer to Appendix A for the Arithmetic Review—particularly the section on positive and negative numbers.) Thus, we now have an equivalent equation: $x + 0 = 3 + 5$ but $x + 0 = x$ and $3 + 5 = 8$ and we finally have a solution: $x = \boxed{8}$. Let us check to make sure our answer is correct. If we substitute 8 for x in the original equation, $x - 5 = 3$, we get $8 - 5 = 3$, which is correct.

   Suppose we had another example: $x + 2 = 7$. Again, we could think of $x + 2$ as being on the left side of the balance and 7 being on the right side, pictured below:

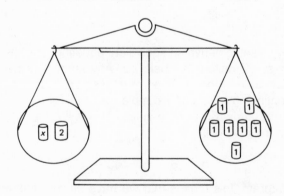

We can also subtract 2 from each side. Again, this will help us to get x all by itself on the left side of the equation. In our picture this will correspond to removing a weight amounting to 2 units from each side. We still have a balance:

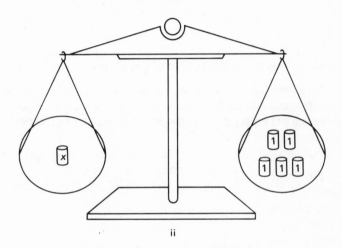

ii

Going back to our equation this means that $x + 2 = 7$ is equivalent to $x + 2 - 2 = 7 - 2$. Again, from steps covered in the Arithmetic Review: $2 - 2 = 0$ and $7 - 2 = 5$. Thus, $x + 2 - 2 = 7 - 2$ is equivalent to $x + 0 = 5$, which is equivalent to $x = 5$, and our equation is solved. Thus, through a comparison of solving equations to maintaining a balance with scales you might see in a biology lab, a butcher's shop, or even as a symbol for a lawyer, we have been able to derive the first major rule to apply to solving basic equations:

Rule 1. If any number is added to or subtracted from both sides of an equation, an equivalent equation results.

Are you finding these beginning examples to be quite easy? Fine!

Remember that developing math skills and confidence depends on attempting and succeeding with basic problems and progressing to more complicated ones. Your success with basic problems will give you the confidence needed to approach more difficult ones. The problems below will give you practice in using your first tools for solving equations:

(1) $x - 7 = 22$
(2) $3 + x = 11$
(3) $5 = y + 2$
(4) $m + 9 = 18$
(5) $2 = 7 - c$
(6) $6 = -5 + t$

Answers are given at the bottom of the page.

Note that in some of the above examples our letter or variable is on the right side of the equation. This means that in our last step we will get an equation of the form $\square = x$, which is certainly just as acceptable as $x = \square$.

Our second rule in solving basic equations is going to allow us to multiply or divide both sides of an equation by the same number to arrive at an equivalent equation. Suppose we had:

$$\frac{x}{2} = 5$$

Look again at our diagram of a balance, or scales.

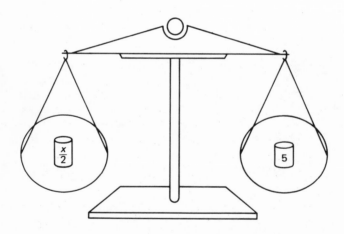

We might think of the quantity or weight $x/2$ being on the left side of the balance and 5 being on the right. Keep in mind that our goal is to get $x = \square$, where $\square$ contains a real number. Suppose we multiply each side by 2. That is, take $2 \cdot x/2$ and $2 \cdot 5$. You may want to refer to the Arithmetic Review in Appendix A once more to recall that twice a number is another way of taking the number and adding it to itself. Twice the number $x/2 = 2 \cdot x/2$, which is really $x/2$ plus another $x/2$. Likewise, twice the number 5 is actually 5 plus 5. We can again picture this process on the scale below, where it is easier to see that a balance will be maintained:

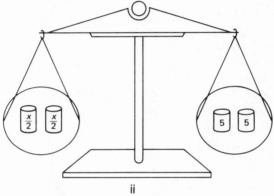

ii

We can see that $x/2 = 5$ is equivalent to $2 \cdot x/2 = 2 \cdot 5$ or $x = 10$.

Division really is the reverse operation to multiplication. The reverse of taking twice a number is to take one-half the number. If I have 10 items, and I double them, I get 20 items. If I start with 20 items and do the reverse operation, take one-half of them (Recall that dividing by 2 is the same as taking ½ of the number or multiplying by ½), I am back where I started with 10 items. Since division is the reverse of multiplication, it can be used in the same way to derive equivalent equations. Look at the next example.

Suppose we have $2x = 18$. Once more we keep in mind that we are striving to obtain $x = \square$. Divide each side of the equation by 2. Thus, $2x = 18$ becomes

$$\frac{2x}{2} = \frac{18}{2}$$

Look at the diagram below to see how this step corresponds:

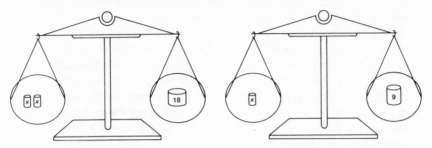

Note that we are again using the idea that $2x$ is made up of x x or $x + x$. Thus, $2x = 18$ is equivalent to $^{2x}/_2 = {^{18}/_2}$, which is equivalent to $x = 9$. Checking, $2(9) = 18$, so our result is correct. These examples give us a procedure for a second rule to be applied to solving equations:

Rule 2. If we multiply or divide two sides of an equation by the same number, we obtain an equivalent equation. That is, this new equation will have the same solution or solutions as the original equation.

Rule 2 actually needs one more statement to make it totally correct. What happens if we take an equation, $2x = 6$, and multiply each side by zero? We will get $0 \cdot 2x = 0 \cdot 6$ or $0 = 0$. (Recall again from Appendix A that 0 times any number is again zero.) We have wiped out our equation entirely. In Rule 2 we must restrict ourselves to multiplying or dividing each side of an equation by a *nonzero* real number. You may want to try your hand at the next few examples, which can be solved by using Rule 2. (Answers can again be found at the bottom of the page.)

(1) $4y = 64$
(2) $6w = 24$
(3) $35 = -7b$
(4) $^t/_5 = 4$
(5) $^x/_7 = 3$
(6) $^a/_4 = 2$

Answers: (1) $y = 16$ (2) $w = 4$ (3) $b = -5$ (4) $t = 20$ (5) $x = 21$ (6) $a = 8$

The next step in our building-block process of solving equations is to apply both rules to slightly more involved equations. Take the following equation:

$$2x - 3 = x + 11.$$

In this equation the variable appears on both sides of the equation. In equations such as the one above you will find it most helpful to use Rule 1 to put all terms containing the variable on one side of the equation and all other terms on the other side of the equation. Remember that our goal in solving the equation is to arrive finally at a step where $x = \square$. Let's return to our example:

$$2x - 3 = x + 11$$

It is customary, but not required, to place the variable terms on the left and the number terms on the right side of the equation. We might begin to solve our equation by applying Rule 1 and adding 3 to each side:

$$2x - 3 = x + 11$$
$$\text{becomes}$$
$$2x - 3 + 3 = x + 11 + 3.$$

Simplifying, we have:

$$2x + 0 = x + 14$$

or

$$2x = x + 14$$

We still need to get all the variable terms on the left side. We again apply Rule 1 by subtracting x from each side:

$$2x - x = x + 14 - x$$

Refer again to Appendix A to recall that $2 + 3 + 5 = 3 + 2 + 5$. In other words, the order in which we add numbers does not change the sum. Thus, $x + 14 - x$ is the same as $x - x + 14$, which is $0 + 14$ or simply 14. Our latest equivalent equation is therefore:

$$2x - x = 14$$

But $2x - x$ is x and we have

$$x = \boxed{14}$$

We can summarize our steps as a series of equivalent equations:

$$2x - 3 = x + 11$$
$$2x - 3 + 3 = x + 11 + 3$$
$$2x + 0 = x + 14$$
$$2x = x + 14$$
$$2x - x = x + 14 - x$$
$$2x - x = x - x + 14$$
$$2x - x = 0 + 14$$
$$2x - x = 14$$
$$x = 14$$

After you have had more practice solving equations you will be able to skip some of the intermediate steps. For example, we might summarize our problem as follows:

$$2x - 3 = x + 11$$
$$2x = x + 14$$
$$x = 14$$

If you think the in-between steps give you a better understanding of the process, continue to include them. Do not feel that you must solve the problem in the smallest number of steps. You want a solution that is logical and that you feel confident with. If writing out each step gives you this confidence, keep on doing it!

Some equations where variables are on both sides of the equation will require using both Rule 1 and Rule 2 to arrive at a solution:

$$6a - 4 = 12 - 2a$$

Again, our goal is to get all of the "a terms" on the left and the number terms on the right. This will allow us to find $a = \square$.

We begin this time by adding $2a$ to each side. Last time we began by adding numerical terms to each side. Feel free to choose either order. Just make sure that you apply Rules 1 and 2 correctly (i.e., whatever you do to one side of the equation you also do to the other). After adding $2a$ to each side of the equation, we now have an equivalent equation:

$$6a - 4 + 2a = 12 - 2a + 2a,$$

which is equivalent to

$$6a + 2a - 4 = 12 + 0,$$

which is again $6a + 2a - 4 = 12.$

At this point we still have number terms on both sides of the equation. Add 4 to each side. We get

$$6a + 2a - 4 + 4 = 12 + 4,$$

or

$$6a + 2a + 0 = 12 + 4.$$

Notice that we now have all the variable terms on the left side of the equation and all number terms on the right side. Now all we have to do is combine *like terms*, that is add, or combine, the variable terms and add, or combine, the numerical terms.

We have: $8a = 16.$

Now we can apply Rule 2 and divide each side by 8. Thus, $8a = 16$ becomes

$$\frac{8a}{8} = \frac{16}{8}$$

or $a = 2.$

Checking, we see that if 2 is substituted for a in the original equation we get:

$$6(2) - 4 = 12 - 2(2)$$

or

$$12 - 4 = 12 - 4$$

or

$$8 = 8,$$

which is a true statement, or mathematical sentence.

Notice that in this example we applied Rule 2 first and then Rule 1.

What would happen if we applied Rule 2 first? Take this equation:

$$3x - 2 = 5$$

Suppose we divide each side by 3. We get

$$\frac{3x - 2}{3} = \frac{5}{3}$$

or $\quad\dfrac{3x}{3} - \dfrac{2}{3} = \dfrac{5}{3}$

or $\quad x - \dfrac{2}{3} = \dfrac{5}{3}$

or $\quad x - \dfrac{2}{3} + \dfrac{2}{3} = \dfrac{5}{3} + \dfrac{2}{3}$

or $\quad x = \dfrac{7}{3}$

We are still able to get the answer, but applying Rule 1 first cuts down on the number of fractions we have to work with. We can now summarize our rules for approaching more complicated equations as Rule 3 below:

Rule 3. Combine like terms and then apply Rule 1. Finally, apply Rule 2.

Let's apply Rule 3 to the equation below:

$$3x + 2 + 5x = 18 + x - 2$$

Combining like terms we get

$$(3x + 5x) + 2 = (18 - 2) + x$$
or $\quad\quad 8x + 2 = 16 + x$

Subtracting 2 from each side we get

$$8x + 2 - 2 = 16 + x - 2$$
or $\quad\quad 8x = 14 + x.$

Now subtracting x from each side we get:

$$8x - x = 14 + x - x$$
or $\quad\quad 7x = 14.$

Note that once you get to a step where you have

$$\square x = \square$$

you can apply Rule 2 to solve for x.

Finally we divide each side by 7 to arrive at

$$\frac{7x}{7} = \frac{14}{7}$$

or $\quad\quad x = \boxed{2}$

Summarizing our steps we have:

$$3x + 2 + 5x = 18 + x - 2$$
$$8x + 2 = 16 + x$$
$$8x = 14 + x$$
$$7x = 14$$
$$x = 2$$

I'll leave the checking to you for this example.

In the next example we give two methods for solving the equation. The point of this example is to demonstrate that there are often various ways one can go about solving a problem, and depending on the problem, one method may be easier than another. Suppose we are asked to solve this equation:

$$6 - 2y = 27 + y$$

We do not have any combining to do on either side so we can go to step 2.

Method I: In most of our previous examples we have tried to place the variable on the left and the number on the right. Following this same procedure we first subtract 6 from each side:

$$6 - 2y = 27 + y$$

becomes

$$6 - 2y - 6 = 27 + y - 6$$
$$6 - 6 - 2y = 27 + y - 6$$
$$0 - 2y = 27 - 6 + y$$
$$-2y = 21 + y$$

Now we subtract y from each side and obtain
$$-2y - y = 21 + y - y$$
$$-3y = 21.$$

Now we are ready to apply the third step of our equation solving procedure:

We divide each side by -3 and get:

$$\frac{-3y}{-3} = \frac{21}{-3}$$

or
$$y = -7$$

(Refer to Appendix A on signed numbers.)

Some adults feel more comfortable working with positive numbers than negative ones. The last few steps in Method I are procedures they may not feel confident with. These adults often choose to apply our equation-solving techniques in such a way that they end up with $\square = x$. Again, this is perfectly acceptable since, for example, $x = 9$ and $9 = x$ are both math sentences saying the same thing. (i.e., if the variable x has the value 9, the original equation is true).

Method 2:

$$6 - 2y = 27 + y$$

If we add $2y$ to each side we will eliminate the $-2y$ on the left and get

$$6 - 2y + 2y = 27 + y + 2y$$
or $$6 + 0 = 27 + 3y$$
or $$6 = 27 + 3y$$

Now we again want the variable term all by itself on one side, in this case the right, and all the number terms on the left. To do this we subtract 27 from each side and obtain

$$6 - 27 = 27 + 3y - 27$$
or $$6 - 27 = 27 - 27 + 3y$$
$$-21 = 0 + 3y$$
$$-21 = 3y$$

Now we are finally ready to divide each side by 3, and we obtain

$$\frac{-21}{3} = \frac{3y}{3}$$
or $$-7 = y$$

In this chapter we have obviously restricted ourselves to certain basic types of equations. There are many other kinds of equations that require other tools to solve them easily, and we will consider them in the chapter on algebra. However, our three rules will allow us to begin solving some basic everyday word problems. We will see in Chapter 4 how our translation and equation-solving techniques become very important steps in forming a plan of action for solving word problems. Before we take an example of how these steps are used, there is one specific type of equation that is worth singling out for special attention.

Suppose we have an equation such as the one below. Here we have an equation of two fractions set equal to each other:

$$\frac{a}{b} = \frac{c}{d}$$

We can apply Rule 2 by multiplying each side of the equation by b:

$$b \cdot \frac{a}{b} = b \cdot \frac{c}{d}$$

or

$$\frac{ba}{b} = \frac{bc}{d}$$

(Refer to Appendix A.)

Simplifying, we get $a = {}^{bc}\!/_d$

We can again apply Rule 2 by multiplying each side by d:

We then have $a \cdot d = {}^{bc}\!/_d \cdot d$

or

$$ad = \frac{bcd}{d}$$

$$ad = bc$$

Thus, ${}^a\!/_b = {}^c\!/_d$ is equivalent to $ad = bc$

You may recall this step as a *cross product*:

$$\frac{a}{b} \diagdown\!\!\!\!\!\diagup \frac{c}{d}$$

Usually, we just memorized this step, but we can now see *why* it works from our material on solving equations. You may want to think of the cross product with the following diagram in mind:

$$\frac{a}{b} = \frac{c}{d} \qquad \text{means} \quad ad = bc$$

We can use this added piece of information as a shortcut for solving particular equations where we have essentially two fractions set equal to each other.

For example, suppose we have this equation:

$$\frac{x}{3} = \frac{4}{6}$$

By taking the cross-product, we can see that this equation is really equivalent to the final equation below:

$$\frac{x}{3} = \frac{4}{6}$$

means $6x = 3 \cdot 4$

or $6x = 12$

Now we apply Rule 2 by dividing each side of the equation by 6, resulting in this:

$$\frac{6x}{6} = \frac{12}{6}$$

or $x = \boxed{2}$

We'll work a lot with these types of equations in the chapter on ratios and proportions. The idea of using the cross product is just an alternate way of solving equations of this type. You may prefer to use only the first two rules. The reason for some of the confusion associated with math is that there are often several methods of obtaining the same answer. Just because your method is different from someone else's does not mean that it is wrong. As long as you can give reasons for each step and can check that your answer is correct, your method is a good one. It works! The problems below will give you practice for the last type of equation:

(1) $\dfrac{x}{8} = \dfrac{1}{2}$

(2) $\dfrac{a}{5} = \dfrac{6}{10}$

(3) $\dfrac{3}{4} = \dfrac{y}{12}$

(4) $\dfrac{15}{6} = \dfrac{10}{x}$

(5) $\dfrac{2}{8} = \dfrac{a}{20}$

Answers are below.

Answers: (1) $x = 4$ (2) $a = 3$ (3) $y = 9$ (4) $x = 4$ (5) $a = 5$

To conclude this chapter we will preview how translation and problem-solving techniques are going to allow us to approach and solve various word problems.

Suppose that in a word problem we come across a statement such as the following:

Five times a number, less 10, is 40

Furthermore, suppose that knowing this number is essential to solving the rest of the problem. Can we determine the number using our new translation and equation-solving tools? First, we try to translate the verbal statement into a mathematical sentence. If we let n denote our unknown number,

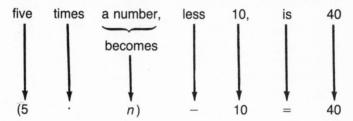

five	times	a number,	less	10,	is	40
		becomes				
(5	·	n)	−	10	=	40

Notice that we group $5 \cdot n$ together since the comma in the verbal phrase implies that we are to take 5 times n and *then* subtract 10.

Now, we have an equation: $5n - 10 = 40$.

All we have to do is solve the equation (i.e., find the value of n, which will make it a true statement):

$$5n - 10 = 40$$

Adding 10 to both sides we get:

$$5n - 10 + 10 = 40 + 10$$
$$\text{or}$$
$$5n = 50$$

Now dividing each side by 5 we obtain the following:

$$\frac{5n}{5} = \frac{50}{5}$$

or $n = \boxed{10}$

Most everyday word problems, when translated into mathematical language, will require that you solve one or more equations. Our skills in translating and solving equations will be essential to our major goal in *Math Without Fear*: applying math to practical, everyday situations. The next chapter gives a guideline to help you form your own organized approach to word problems. Its goal will be to help you get started and stay with the problem.

The exercises that follow are a collection of various equations similar to our examples. You will want to spend time getting practice with these equations before progressing to Chapter 4.

PRACTICE SET

Solve the following equations for the unknown quantity:

(1) $8 - a = 6$

(2) $9s = 360$

(3) $7a = 14$

(4) $-3a = 12$

(5) $\dfrac{2y}{3} = 10$

(6) $2x + 5x = 56$

(7) $5 = (x/_3) + 2$

(8) $(x/_2) - 5 = 11$

(9) $8a - 1 = 47$

(10) $\dfrac{1}{8} = \dfrac{3}{x}$

(11) $\dfrac{5}{12} = \dfrac{x}{60}$

(12) $8 = \dfrac{y}{2}$

(13) $\dfrac{x}{-2} = -7$

(14) $9 + 2x = 5x + 3$

(15) $6y - 4 = y - 24$

(16) $3w + 9w - 7 = 5w$

(17) $8s - 2s + 6 = 4s + 18$

(18) $13x + 14 = 11x + 34 - 3x$

(19) $b - 2b + 6 = 4b - 7b + 20$

(20) $3x + 15 = 5x - 7$

Answers to the problem set on page 66.

Answers: (1) $a = 2$,   (2) $s = 40$ (3) $a = 2$

(4) $a = -4$ (5) $y = 15$ (6) $x = 8$ (7) $x = 9$

(8) $x = 32$ (9) $a = 6$ (10) $x = 24$ (11) $x = 25$

(12) $y = 16$ (13) $x = 14$ (14) $x = 2$ (15) $y = -4$

(16) $w = 1$ (17) $s = 6$ (18) $x = 4$ (19) $b = 7$

(20) $x = 11$

CHAPTER 4
GETTING STARTED AND STAYING WITH IT– PRACTICAL MATH FOR EVERYDAY PROBLEMS

The major goal of this book and the Math Without Fear program is to help math-anxious adults develop math skills that they can apply to daily situations while acquiring a more confident attitude toward math in general.

Chapter 2 stated that the key to achieving this goal is the word problem, or everyday problem. The skills needed to solve word problems are some of the same skills needed to apply math while on the job, shopping, reading the newspaper, or taking a math-related course.

Solving word problems requires a variety of skills, including reading for understanding, determining what questions need to be answered, seeking out unknown quantities, representing verbal relationships by mathematical ones, solving equations and other technical math problems, and finally interpreting your answer in terms of the original verbal problem. No wonder word problems cause so much anxiety! Word problems involve such a variety of different skills. There is no one special formula, no quick-and-easy answer, no one procedure that solves all word problems. However, this doesn't mean that you have to approach each word problem in a haphazard way, hoping that the answer will somehow evolve. You can develop a more organized approach to problems. Those skills mentioned previously can be developed, and through practice you can become more comfortable with, and adept at doing, everyday math problems. Remember—*learning* math means *doing* math.

We began our approach to word problems in Chapter 2 by building a math vocabulary and relearning how to converse in mathematics. We began to translate words into math symbols and then short phrases into mathematical relationships. In Chapter 3 we continued to increase our skills in the language of mathematics by translating verbal sentences into mathematical equations. We saw that an integral step in doing word problems would be solving equations. We introduced a basic math skill: equation solving. Again, through lots of practice, doing more and more problems, we have been able to solve different types of equations. Now we are ready to take the next step: creating a method of approaching everyday problems that works for you.

The method that follows should serve only as a guide for you to build your own problem-solving procedure. You may want to take shortcuts, eliminate steps, or make up steps of your own. One of the major benefits of the following procedure is that it will help you *get started.* Many math-anxious adults just don't know where to begin, where to start. They have no idea of how to go about searching for the answer. Furthermore, this floundering reinforces their fears that they are too dumb to do math. Surely, everyone else could see immediately how the answer is to be determined.

A mathematician, for example, would be able to calculate the answer in a few, quick steps. This picture of the mathematician quickly solving problems could not be further from the truth. Mathematicians often have little or no idea of how to solve a problem when they first read it. However, they do have confidence in their own ability and lots of experience in doing word problems, which allows them to organize their thoughts and to begin to derive an answer.

Rarely does an answer to a word problem pop immediately to mind. Don't let the fact that you have no idea how to solve a word problem upset you. Try not to be overwhelmed by the problem. There are procedures you can apply to give you *control* over the problem. Perhaps you may not be able to get the answer on your first step. However, you will be able to make some conclusions, and you will probably be able to rule out various incorrect answers.

Getting started toward a solution always is the hardest part.

The method that follows will help you take that vital first step. It will allow you to analyze the problem, organize your thoughts and conclusions, and hopefully arrive at an answer.

The translation and equation-solving skills you have been building in Chapter 2 and Chapter 3 will be extremely helpful in this chapter. You will begin by applying your problem-solving method to some typical, everyday problems. Then you will practice with various problems. In doing these problems you will be able to organize and refine your own method for doing word problems. Our procedure is only a guide to help you develop your own approach. As you do more and more problems, your confidence in your own ability will increase and you will begin to gain control over math problems instead of being overwhelmed and defeated by seemingly unanswerable problems.

Let's look at the following word problem:

Example 1. Anita earned $120 more than Charlotte last month. If their total earnings were $1,490, how much did each woman make?

This problem is a typical example of how math is used in everyday situations. A classic case of comparing salaries, it will be a nice problem through which we can define our problem solving procedure and show how to apply it.

WORD PROBLEM PLAN OF ACTION

Our problem-solving method, which we will call our plan of action, consists of the following five steps:

PLAN OF ACTION

Step 1   *READ AND REPRESENT*
Step 2   *GUESS*
Step 3   *TRANSLATE*
Step 4   *SOLVE*
Step 5   *CHECK AND CONCLUDE*[1]

Step 1   Read and Represent

Unfortunately, our first step in beginning word problems does not call for rest and relaxation. However, both of these may result as math confidence builds and success with applying math frees you from previous anxiety and failure. *R and R* in our plan of action stands for *read* and *represent*. This initial step is the most important in building a problem-solving method. In this step you will be discovering what quantities need to be found, what questions need to be answered, and what direction must be followed in arriving at a correct solution. As we have mentioned many times before, the initial reaction of the math-anxious adult to a first reading of a problem like our example is to freeze! There is no way they can answer that problem!

After reading the problem once they are ready to give up. Frustration and a complete lack of confidence have overwhelmed them. One reading of the problem has convinced them that they are incapable of solving it; the problem is confusing, and they don't know where to begin. They assume they are too stupid to get it. Yet these same adults will read a chapter of Plato or Freud, find it also confusing, but think nothing of it. Plato and Freud are supposed to be confusing the first time you read them. However, when you reread the chapter a few times you get a better understanding and begin to appreciate some of the major elements in the reading. You begin to analyze the author's remarks and reword them in your own phrases.

I maintain that these same techniques can be applied to doing word problems. You don't consider yourself stupid just because it took you two or three readings to comprehend an article in philosophy. Many times you learn or understand a little more with each reading. You pick up certain phrases and points you missed before. The same idea can be applied to word problems.

Read the problem once just to acquaint yourself with the subject matter. At this point we might note that our example has to do with salaries and that it involves some kind of comparison. Now read the problem a second time, with paper and pencil at hand, but much more slowly. Reading mathematical problems or statements is not at all like reading a mystery, although you may find it just as puzzling. Every word and symbol is important, and material is much more concentrated. You need to read more slowly. Use a paper and pencil to jot down important ideas and terms in the problem. Using paper and pencil while reading is certainly not restricted to math problems.

My freshman English teacher in college advised his students always to read any literature they wanted to analyze with pencil in hand. In this way they could note important ideas and concepts they might otherwise forget after completing the reading. Many of us have used this technique in various subjects without thinking. Making notes along margins and underlining have always been popular methods for highlighting important concepts. Now there are transparent pink or yellow markers to achieve the same effect. Feel free to use these techniques also. In rereading our particular problem we might jot down the items that we are dealing with:

Anita's monthly salary
Charlotte's monthly salary
the fact that Anita earned more than Charlotte
the fact that their combined salaries equaled $1,490.

This first reading should give you a general idea of the problem. For example, we read "Anita earned $120 more than Charlotte" as "Anita earned more money than Charlotte." Don't worry about the numbers; we'll read it again after we know what kind of problem we have to work with and what solving it will involve.

You want to determine which elements are known and which quantities you have to find. What are the relationships between the various quantities? You may want to underline important words and symbols. Recall from the previous chapter that a symbol can be associated with various English words. For example, X sometimes indicates multiplication and sometimes represents an unknown. Be sure to read each word and symbol carefully.

In a third reading of the problem you will want to determine the main question of the problem. What are you being asked to find? Key phrases in identifying the question will be phrases such as "Find the following . . ." and "What is or are . . . ?" and "Calculate or determine these . . ."

Identifying the question that is being asked may well be your most important step.

In our problem we are asked to find Charlotte's salary; the second part of Step 1 is to represent any unknown quantities. In our problem we have two unknown elements: Anita's salary and

Charlotte's salary. As we did in our translation chapters, we will use letters to stand for these unknown elements. Again, many in the MWF program found it helpful to use letters related to the item they are representing. In our example we might use the letters a and c:

$$a = \text{Anita's salary}$$
$$c = \text{Charlotte's salary}$$

Step 2 Guess

The second step in our plan of action involves making an estimate of the answer. Many math-anxious adults found it useful to make an educated guess at the possible answer, drawing upon previous experience, practice with similar problems, and even intuition. Don't discard your feelings or intuition about a problem. Many math-anxious adults think that intuition and estimates have no place in mathematics. They think that only specific equations, calculations, and exact solutions can be used in the problem-solving process.

This traditional view of mathematics is another math myth. New thoughts or concepts in math have always been created from someone's feelings or intuition about a problem. All mathematicians and scientists begin a project or problem by using intuition. They estimate what the answer should be or how it could be found. Then through various stages of trial and error they go about the process of proving their claim. Various attempts may result in dead ends. Estimates may have to be revised and the whole process repeated.

The development of all mathematics has been due to the intuition and guesses of various math-interested individuals. So do not be afraid to use your *own* judgment and feelings about a problem. Making a reasonable guess about the answer will have various benefits.

First, guessing at the answer will give you a means of getting started with a problem, the hardest step in doing math problems. Particularly for the math-anxious, who lack confidence in their own ability and see finding the solution as an unfruitful test of their nerves, beginning a math problem can represent the start of an unpleasant task. Guessing can help these adults, and you, overcome this initial hurdle.

Second, guessing gives you a sense of control over the problem. You have what you feel is a reasonable estimate of the answer. You have a solution in mind; the problem is no longer a completely baffling puzzle for you. Sometimes your estimate will be sufficient for the answer to the problem. Certainly, if you are comparing the cost of two cars, one of which is $4,025 and the other is $6,987, estimating that the second car costs almost $3,000 more than the first will probably be sufficient to make a financial decision. Knowing that the second car costs exactly $2,962 more may not be necessary at all.

Third, making an educated guess will prevent you from failing to recognize unreasonable conclusions and major errors. For example, if you are comparing prices on cans of vegetables selling for under a dollar, it is quite possible to make an error using decimals or fractions and end up with an answer such as this: two cans of Brand X cost $22.37 more than one large can of Brand Y. Making a reasonable guess, based perhaps on previous shopping experience, will give you at least one check on your answers and conclusions.

In our particular example we might note that Anita and Charlotte together earned about $1,500 and Anita earned about $100 more than Charlotte. If they had earned equal salaries, we would divide $1,500 in half and get $750 each. There is nothing wrong in stopping right here for your estimate, since $750 apiece certainly gives you a definite range for your answer. If you want to go a step further, you might increase Anita's salary by $50 and decrease Charlotte's salary by $50. This will give you the $100 difference in salaries, yet the total will still be $1,500. We might estimate Anita's salary to be $800 and Charlotte's to be $700.

I think it's necessary to point out that this step is a very subjective one and will vary from individual to individual. In MWF it was often surprising how many different ways of thinking about a problem there were. However, this step is consistent with our entire procedure. You want to develop a problem-solving method that works for you, one that makes sense to you. Your estimate should be one that *you* find reasonable or intuitive. It's also necessary to note that this step is not always possible. Solving a problem may involve discovering relationships that you could not have imagined at the beginning of the problem. Don't worry about these cases. All of our recommendations for forming your own plan of action are just that, *recommendations.* You may find some of these tips to be of little help in solving some problems, while others work well in almost every problem you try. No one strategy or formula works for all problems. However, we found that developing an organized approach, where some or all of the given steps are used, can help most math-anxious adults begin to work with everyday problems.

Step 3   Translate

In this step we translate the verbal relationships from Step 1 into mathematical ones. You will be applying the skills you have developed from Chapter 2 and Chapter 3. Let's go back to our original example.

In Step 1 we noted that there were really two different relationships given or implied in the problem. First of all, we are told that Anita earned $120 more than Charlotte did last month. That is, Anita's salary is $120 more than Charlotte's. Keeping in mind that we are letting a represent Anita's salary and c represent Charlotte's salary, we get the following equation:

$$a = c + 120$$

Second, their combined salaries are $1,490. This means that Anita's salary plus Charlotte's salary is $1,490. Translating, we get $a + c = \$1,490$. Recalling our work in problem solving from Chapter 3, we know that if an equation involves only one letter or variable, we can solve it. We note that Anita's salary is really just Charlotte's salary plus $120. Thus, in our last equation, we can make this substitution, and we get this:

$$(c + 120) + (c) = \$1,490$$
$$\text{(Anita's)} \qquad \text{(Charlotte's)}$$

Step 4   Solve

In this step, we are going to do the calculations most of us equate with doing "real math"—equation solving. Using the skills we learned in Chapter 3, we try to find the number that, when substituted for c, gives us a true equation. Our first step is to combine the c terms:

$$c + 120 + c = 1,490$$

which becomes

$(c + c) + 120 = 1,490$
$2c + 120 = 1,490$ (Now we want to isolate the c term):
$2c + 120 - 120 = 1,490 - 120$
$2c = 1,370$
$\dfrac{2c}{2} = \dfrac{1,370}{2}$
$c = 685$

Note that 685 gives us Charlotte's salary. Was this all the problem asked for? If we go back to the original problem, we see that we are asked to find *both* Anita's and Charlotte's salaries. There is a natural tendency for all of us to stop when we reach $c = 685$. This is why it's so important to determine exactly what we are to find in Step 1.

Anita's salary is Charlotte's plus 120. Thus,
$a = c + 120 = 685 + 120$, or 805.

Step 5   Check and Conclude

In this step we want to check our final solution against our estimate and other relationships given in the problem. We might note that 805 and 685 do give a total of 1,490.

That is: $805 + 685 = 1,490.$

Second, we might also note that our final answer certainly seems reasonable when compared with our estimates of 800 and 700.

Finally, it's necessary to make a final conclusion with respect to the original problem. The numbers 805 and 685 have no meaning unless we make the final interpretation where the answer to the question, "What salaries did Anita and Charlotte receive last month," is this: "Charlotte received $685, and Anita received $805."

A recapitulation of our word problem plan of action follows:

PLAN OF ACTION
Step 1   READ AND REPRESENT
Step 2   GUESS
Step 3   TRANSLATE
Step 4   SOLVE
Step 5   CHECK AND CONCLUDE

The rest of this chapter is devoted to approaching and solving everyday word problems. I have tried to include a variety of basic problems that you might encounter while shopping, on the job, or even reading the newspaper.

The rest of this book will introduce different math topics in relation to given word problems.

For example, percentages will be introduced as a means of solving consumer problems.

We begin by looking at some basic word problems and building up to more difficult ones. You would not begin to learn to play the piano with Chopin! In the same respect we are not going to begin by tackling a calculus problem. Developing math skills is a progression from simple skills to more complicated ones. Each step in acquiring math skills and confidence requires building on the skills from previous steps.

Similarly, math confidence will build as each successfully solved problem gives you more confidence to go one step further.

This building-block progression naturally will involve lots of *practice*. Many adults think that if they know how to solve equations, then they should be able to do word problems. No wonder they become confused and anxious when their equation techniques do not seem to be enough!

We have already seen that solving word problems requires a variety of skills, both verbal and mathematical, some intuition, translation skills, and even common sense. In addition, there are many different kinds of word problems, each with its own characteristics. How does one develop the skills necessary to succeed with all these problems?

Practice!

You can become adept at word problems only by doing lots and lots of problems. Then when you try to solve a new problem that is similar to one you have solved before, the solution will be that much easier. You will develop a way of organizing your work and approaching the problem. Having done several problems of a similar kind will be essential to solving the problem confidently. In other words, I am promoting the same technique that you have used in probably every other area of your life. Skill and ability come from *practice* and *perseverance*.

That first time you drove a car, the first time you tried to ski down the beginner's hill, the first time you cooked an entire meal—these initial attempts probably did not result in your immediately being classified as a great driver, skier, or cook! However, you did not give up and say, "I never could ski" or "Driving is not for me" or "I just don't have a gourmet mind!" You may still not be a *great* driver, skier, or cook. However, you probably have been able to acquire a certain adeptness by getting practice, making mistakes, and learning from each experience. The same will hold true with math.

Do not let mistakes or errors discourage you. Learning any skill or new topic will involve making errors. If you allow difficulties to discourage you, confidence and skill can hardly be acquired.

Learning to do word problems is much like sticking to a diet. Day one is terrible. But if you make it through day one, day two is that much easier. The idea is that if you succeeded yesterday, you probably have the ability to get through today also.

Similarly, that first word problem seems so difficult. With each successful conclusion of a word problem, however, you build a more positive and confident attitude toward math in general. One of the most pleasurable aspects of the MWF program was to see adults who would normally cringe and give up at the sight of some word problem suddenly see a solution begin to evolve. Especially rewarding to these adults was the fact that they were obtaining these solutions on their own. Throughout the rest of this chapter and the rest of this book I will be giving helpful hints that seemed to work well with most of the members in the program.

The next problem is typical of a whole class of problems about relationships between numbers.

More Examples and Helpful Hints

Example 2:   The sum of 2 numbers is 24. One number is four more than the other. Find the two numbers.

Step 1   Read and Represent

In reading and rereading this problem we see that we are to determine two numbers related in some way. The question is to find the values of these two numbers.

Helpful hint: Underline the question in the problem.

Helpful hint: Rewrite verbal relationships in your own words. In our problem, we might rewrite the first relationship as follows: The first number plus the second number is 24. We might rewrite the second relationship as follows: The second number is the first number plus 4.

Helpful hint: Don't be afraid to write notes in English. You are probably more comfortable with verbal expressions than with mathematical ones, and the process of writing your first observations will help you get a clearer picture of the problem.

To finish Step 1, we must represent our two unknown numbers. If the first number is denoted by m, then the second number, which is the first number plus 4, becomes $m + 4$.

Step 2   Guess

Given that our two numbers differ only by 4, we might estimate that our two numbers are almost the same. We could approximate them by guessing each to be about one-half of 24, or 12.

Step 3   Translate

We note that we have already translated the verbal relationship that the second number is the first number plus 4. We have only to translate "the first number plus the second number is 24." Recall that this is our rewritten expression from Step 1. Translating we get:

The first number plus the second number becomes 24
$$m + (m + 4) = 24$$

Helpful hint: It is sometimes helpful to put parentheses around terms involving more than one letter or number. This will save some confusion in more complicated problems.

Step 4   Solve

Note that $m + (m + 4) = 24$ is an equation in one letter or variable and can be solved by methods discussed in Chapter 3.

Thus, $m + (m + 4) = 24$ becomes
$$m + m + 4 = 24$$
$$2m + 4 = 24$$
$$2m + 4 - 4 = 24 - 4$$

$$2m = 20$$
$$\frac{2m}{2} = \frac{20}{2}$$
$$m = 10 \text{ and } m + 4 = 10 + 4 = 14$$

Step 5   Check and Conclude

Indeed, 14 is 4 more than 10. Also, the sum of our two numbers is 10 + 14, or 24. Our answer meets all of the requirements of the problem, and we can conclude that our two numbers are 10 and 14.

Example 2 is typical of a whole class of problems involving relationships between numbers. In some cases, the actual numbers may represent salaries, as in Example 1, or they may denote costs, numbers of objects, or measurements.

Helpful hint: For some problems, many in MWF found it helpful to draw a picture or diagram to represent the problem. The following is such a problem. You will also note that this prob- lem requires some familiarity with cars.

Example 3: Because of engineering requirements, an automotive manufacturer makes vans with the following dimensions:

Front bumper to front axle is 60 cm.
Front axle to rear axle is 4 cm. more than three times the distance from rear axle to rear bumper.
The overall length is 344 cm.*
Find the distance from the rear axle to the rear bumper. Also, find the distance between axles.

I admit that at first glance this problem gave me a few mo- ments of anxiety. Most of us in the MWF program felt lacking when it came to cars and how they worked. Many of us were not even sure where the axles are in a car. Before beginning to solve the problem, we needed to refer to a diagram of the basic structure of axles and bumpers in the average car:

---

*Recall that cm. stands for centimeters in the metric system.

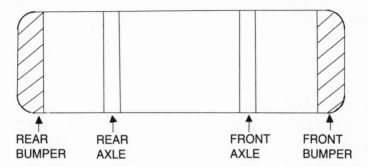

Helpful hint: In following our basic 5-step procedure we use the above diagram to help represent the various quantities involved.

Step 1   Read and Represent

Upon rereading the problem, we see that there are actually four distances involved:

(1) the overall length of 344 cm.

(2) the 60 cm. from front bumper to front axle

(3) the unknown distance from rear bumper to rear axle (Let's use *d* to denote this distance.)

(4) the unknown distance from front axle to rear axle

We note that this distance is 4 cm. more than 3 times the distance from rear axle to rear bumper. We might rewrite this quantity as 3 times *d* plus 4, or $3d + 4$.

Now we denote these distances on our diagram.

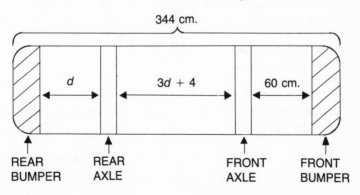

Step 2   Guess

Using my inadequate knowledge of cars, I was tempted to guess that the distance from rear axle to rear bumper was approximately the same as the distance from front axle to front bumper. Cars seemed to be, on the average, as long in the front as in the back. I guessed that d was about 60 cm. Again, my estimate is far from accurate, but it will help to prevent me from making errors such as d = 300 cm.

Step 3   Translate

We already have represented all of our unknown quantities. What is the only relationship tying all the distances together? This relationship is implied and probably requires a couple of readings to determine. Let us look for a quantity or expression relating to all of the distances. It's the overall distance of 344 cm. In other words, 344 cm. is the distance from front bumper to front axle plus the distance from front axle to rear axle plus the distance from rear axle to rear bumper. Translating we get this equation:

$$344 = 60 + (3d + 4) + d$$

(Note that we have again used parentheses to separate the second distance.)

Step 4   Solve

$$344 = 60 + 3d + 4 + d$$
$$344 = (60 + 4) + (3d + d)$$
$$344 = 64 + 4d$$
$$344 - 64 = -64 + 64 + 4d$$
$$280 = 4d$$
$$\frac{280}{4} = \frac{4d}{4}$$
$$70 = d$$

Also, the distance from front axle to rear axle = $3d + 4$ becomes $3(70) + 4$

= 210 + 4, or 214

You will also note that these distances should be expressed in centimeters.

Step 5   Check and Conclude

Using our answers from Step 4, we can now fill in our diagram as below:

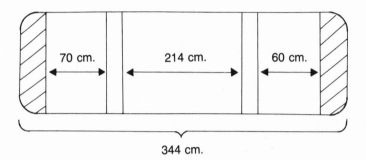

344 cm.

Our answer, d = 70, is certainly close to our estimate of 60 cm. We might check further to conclude that 70 + 214 + 60 = 344. In our final conclusion we must be careful to see that we have answered the questions given: 214 cm. is the distance from front axle to rear axle, and 70 cm. is the distance from rear axle to rear bumper.

Helpful hint: Using a diagram to display unknown quantities and final answers will prevent you from confusing answers. For example, it will help prevent you from putting 214 cm. for the smaller distance and 70 cm. for the larger distance.

Example 4: Helen has just started taking a continuing education class in stained glass artistry. As part of her first project she is given a piece of lead 25 inches long. She needs to break this piece of lead into 3 pieces. Two pieces are to be the same length and the third is to be three times as long as the first two. How should she break up the given piece of lead to satisfy the requirements?

Step 1   Read and Represent

In this problem the unknown quantities are the lengths of the three pieces of lead that are to be cut from the original piece. Let us represent the lengths of the two shorter pieces by L. The length of the third piece is 3 times the length of the other two. Thus, we can denote this length by $3 \cdot L$ or $3L$. Note that since the 3 new pieces were cut from the original piece, the sum of the lengths of these 3 pieces should be 25 inches. This relationship could be depicted in the diagram below.

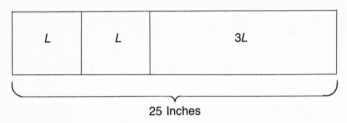

25 Inches

Step 2   Guess

In this problem we might note that the two shorter pieces equal a little less than half of the original piece. The longest piece makes up a little more than half of the original piece. If we think of the two shorter pieces as half of 25 inches, we might conclude that the 2 pieces together are about 12 inches long. This means that each would be 6 inches in length. Again, estimating, we might say that the third piece is about half of 25, or maybe 13 inches long. These estimates give us a reasonable range for the answer.

Step 3   Translate

In order to find L we need to find some relationship involving the two shorter pieces of lead and the longer piece. We have already mentioned this relationship: all of the lengths put together equal 25 inches. That is, the sum of the lengths of the two shorter pieces plus the third longer piece is 25 inches.

Translating we obtain: $L + L + (3L) = 25$

Step 4   Solve

$$L + L + 3L = 25$$
$$2L + 3L = 25$$
$$5L = 25$$
$$\frac{5L}{5} = \frac{25}{5}$$
$$L = 5$$

Step 5 Check and Conclude

If L = 5 is the length of the two shorter pieces, then $3L$ = 3 · 5, or 15 inches, is the length of the longer piece. We might check that their sum, 5 + 5 + 15, is indeed equal to 25. Therefore, our conclusion is that Helen should cut her original piece of lead into 3 pieces measuring 5, 5, and 15 inches to meet her requirements.

Our last major example for this chapter is a word problem typical of those encountered by many adults in jobs concerned with sales.

Example 5: Lynn sells Oriental rugs. Her weekly salary is $140, plus $80 for each rug she sells. During the week before New Year's Day she made $540. How many rugs did she sell?

Step 1 Read and Represent

Our unknown quantity in this example is the number of rugs, which we could denote by R. Rereading the problem, we might note that R must be related to the $540 income in some way. At this point you will want to use your experience as a consumer to discover this relationship. If she sold one rug, what would her income be for the week? She would have $140 plus $80. What if she sold 2 rugs? Her income would then be $140 plus $80 for each rug, or $140 plus 2 times $80, or $140 plus $160. If she sold 3 rugs, her income would be $140 plus 3 times $80.

Helpful hint: If a relationship between two quantities seems to be eluding you, take the simplest example you can think of and keep taking more examples. This process does not *always* result in the desired expression, but it will always give you a better feel for the problem. Also, concrete examples are easier to work with before getting a general solution. You are more comfortable with these cases, and you will feel more in control of the problem.

Therefore, we see a pattern developing, and we might see that if Lynn sells R rugs, she would make $140 plus R times $80.

Step 2 Guess

At this point you might note that only $540–$140 = $400 really concerns income from the number of rugs sold. The $140 is a flat salary. Then we might note that if Lynn sells 10 rugs, she will make an additional 10 times $80 or $800. This figure is, of course, too large, but it does tell us that Lynn sold fewer than 10 rugs that week.

Step 3   Translate

In Step 1 we discovered that Lynn's income, $540, was $140 plus R times $80. This translates into the following:

$$\text{Income} \quad \$540 = \$140 + (R \cdot 80)$$

Step 4   Solve

$$540 = 140 + R \cdot 80$$

becomes

$$540 - 140 = 140 + (R \cdot 80) - 140$$
$$400 = R \cdot 80$$
$$\frac{400}{80} = \frac{R \cdot 80}{80}$$
$$5 = R$$

Step 5   Check and Conclude

Five rugs certainly appears to be a reasonable answer. In fact, if we check her income, $140 plus 5 times $80, we obtain $140 + (5 \cdot 80)$, or $140 + $400 = $540, the actual income. We can conclude that Lynn must have sold 5 rugs during the week before New Year's.

In this chapter we have talked about word problems in general—what kinds of tools and techniques are needed to solve word problems, how to go about approaching word problems and developing your own problem-solving method. We have also included helpful hints to get you started and keep you going. In addition, we worked with different examples of basic word problems.

The remaining chapters of *Math Without Fear* are also devoted to the word problem and how to increase your success in problem solving while reducing your math anxieties. Further tips on relieving fears and developing various math skills are included. New math skills are introduced as they are needed in various types of word problems. You have probably noticed that the math problems we have begun working with seem to make more sense, and procedures are not as vague as when you studied math previously.

As you practice with the examples and the problems at the end of this chapter, you should begin to feel at least more comfortable, if not quite confident about approaching math in general. As you continue through the rest of *Math Without Fear* try to apply new materials to problems you may have met during the day. For example, you will probably want to apply materials from the chapter on ratios to grocery shopping. Remember that in order to develop math skills you are going to have to attempt more and more math-related problems. Try to note your reactions to math-related day-to-day problems—figuring out tips, doing comparison shopping, computing taxes—as you progress through the book. The more contact you have with math, the more comfortable you will feel with it.

In the next chapter we apply our new problem-solving skills to ratios, or making comparisons. For example, if 3 cans of beans cost 91¢, how much will 7 cans cost?

The following group of problems will give you practice in working with different everyday word problems:

PRACTICE PROBLEMS

(1) After gaining 10 pounds, Andrea weighed 133 pounds. What was her original weight?
(2) The cost of a dozen eggs is 89¢. This price is an 8¢ increase from 2 months ago. What was the previous price?
(3) A major sports event was attended by 124,000 people. This was 15,000 more than had attended the same event last year. What was last year's attendance?
(4) After using a wood stove for a year, John sold it for $225 less than the amount he paid for it. If the amount he received was $275, how much did he pay for it originally?
(5) A woman sets out to drive from Scranton to Syracuse, a distance of about 295 miles. If she drives 150 miles, how many miles is she from Syracuse?

(6) The census bureau noted the number of children belonging to four different families in a rural town in Pennsylvania. Collectively, there were 14 children. Individually, families No. 1 and No. 2 had the same number of children. Family No. 3 had one less child than family No. 1, and family No. 4 had 2 times as many children as family No. 1. How many children did each family have?

(7) The sum of Edith's age and Harriet's age is 94. If Edith is 51, how old is Harriet?

(8) There is a number such that 2 times the number, plus 7, is equal to 43. Find the number.

(9) The sum of two numbers is 11. The larger number is five less than three times the smaller number. Find each number.

(10) Richard has a budget of $540. If he spends twice as much on food as clothing and three times as much for rent as food, how much does he spend for each?

(11) While baking a cake you notice that the cake together with the icing requires 24 ounces of butter. The cake itself requires twice as much butter as the icing. How many ounces of butter are needed for the icing?

(12) Suppose you burn up three times as many calories while walking as you do while sitting. Furthermore, suppose you spent eight hours sitting and one hour walking on your day off from work, and these activities burned up 1,100 calories. How many calories did you burn up walking?

(13) A 30-foot piece of copper tubing is to be cut into 3 pieces. One piece is to be 7 inches long. Of the two remaining pieces, one is 5 inches longer than the other. What are the lengths of the 3 pieces?

(14) In a chorus line it is important for all of the dancers to appear to be the same height. This is sometimes achieved by having the dancers wear hats of varying heights. Maryanne and Jane are 62 and 64 inches tall, respectively. With their hats on, they are the same height. If Maryanne's hat is twice as high as Jane's, what is the height of Jane's hat?

(See Appendix B for answers)

REFERENCES

1. Selby, Peter H. *Practical Algebra*. New York: John Wiley & Sons, 1974.

CHAPTER 5
MAKING COMPARISONS– RATIOS AND PROPORTIONS

If 3 yards of fabric cost $4.80, how much will 5 yards cost?

A giant-size box (13 oz.) of cornflakes costs 91¢. The regular size (10 oz.) costs 69¢. Which size is the better buy?

If a car travels 138 miles on 6 gallons of gasoline, how many miles will it travel on 8 gallons?

RATIOS

Questions like the ones above are typical of some of the comparisons and decisions we are forced to make each day. In these days of uncertain employment, ever increasing inflation, and the constant threat of oil embargoes and gas shortages, activities such as determining gas mileage and shopping comparatively can be extremely important. Decisions made regarding these activities can be vital to your future financial security, retirement, education plans, and even recreational activities, such as trips and hobbies.

The fact that these decisions involve applying math skills effectively probably is not new to you. Gas shortages and inflation are great sources of anxiety in themselves. To have this anxiety compounded by your fear of mathematics eliminates most chances of making an effective and reasonable decision. Your emotions and fears probably will keep you from thinking clearly and will lead to confusion, resulting in some random choice. It seems that inflation and the financial crisis may be with us for some time, as well as the anxieties associated with them. Dispelling these financial worries is a task that I am afraid even our economic advisers have not been able to accomplish. However, we can reduce the anxieties that accompany the math skills involved in these financial comparisons. Comparisons, or ratios, as we will refer to them later, will be one key to becoming effective consumers.

In MWF, ratios became a topic especially enjoyed by most of the participants. Everyone seemed to do well, not only in solving comparison problems but in formulating and answering such problems of their own. I think that one reason for the interest and success in this area is the great deal of experience every adult has with being a consumer. Whether grocery shopping, buying a car, or comparing prices in discount drug stores, making any purchase involves making comparisons. Unless you have more money than you know what to do with and price is no object, you have been using mathematics to make your decisions, whether you have realized it or not.

A major goal of this chapter is to give you more confidence and better organization in approaching these financial decisions. You will not only gain control over a vast area of math-related problems, but you should also develop more control over your consumer affairs.

Before we begin to solve problems, we should say a few words about ratios or comparisons and how they are used. The creation of numbers was no doubt a result of people's need to count. Another cause of the development of numbers was probably the desire to make comparisons. For example, one often needs to make comparisons between prices while shopping for an article. Children rely on numbers to compare the number of toys they have to the number their friends have. We adults often draw comparisons between one another's salaries. Comparison is one of the most basic and essential math tools we use in daily situations.

Mathematically there are two methods of making comparisons. If I ask, "How does a nickel compare with a quarter?" you might answer in one of two ways. First, you could tell me that a nickel is worth 20¢ less than a quarter. Second, you could state that the nickel is worth 1/5 as much as the quarter. This alternate reply is obtained by dividing 5¢ by 25¢: $5¢/25¢ = 5/25 = 1/5$. This result, 1/5, is the *ratio* of a nickel to a quarter. To be explicit, a *ratio* of two quantities is the first quantity divided by the second quantity. In other words ratios are fractions. Ratios may be expressed by the word *per*, as in miles per hour or revolutions per second, and are written $^{miles}/_{hour,}$ $^{revs}/_{sec.}$ Let us look at another example.

Suppose we are comparing the number of men to the number of women in a Math Without Fear workshop. Furthermore, let us suppose that there are 10 men and 20 women. The ratio of men to women would be $^{10}/_{20}$. There are various ways to express ratios. Our given ratio, $^{10}/_{20}$, could also be expressed as 10 : 20, or 10 ÷ 20, or ½. In this last representation, the ratio, or fraction, has been reduced from $^{10}/_{20}$ to ½. These fractions are equivalent: the ratio 10 : 20 is equivalent to 1 : 2. In words, we might say that for each man there are two women.

Ratios can be stated in various ways. In the next example we see some of these variations:

Example 1: Suppose a weight-watchers group has a total of 28 people in it. There are 21 females and 7 males. (a) What is the ratio of females to males? (b) What is the ratio of males to females? (c) What is the ratio of females to the total number of people in the group?

(a) The ratio of females to males means the number of females divided by the number of males, which give us $^{21}/_7$ = $^3/_1$. So the ratio of females to males is 3 : 1.

(b) This is just the reverse of part (a). The ratio of males to females is 1 : 3.

(c) Here we have 21 females out of a total of 28 people. We note that $^{21}/_{28}$ = $^3/_4$, and therefore our ratio is 3 : 4.

The next example should be familiar to any seasoned grocery shopper.

Example 2: While rushing down the aisle at a local grocery store near closing time, I realized that I had forgotten to buy corn- flakes for the next morning. Not to let haste prevent me from being economical, I began to compare the various sizes of my cereal. The giant-size box (13 oz.) sold for 91¢, while the regular-size box (10 oz.) was selling for 69¢. In order to make a comparison, it was necessary for me to form ratios

$$= \frac{\text{cost of the box of cereal}}{\text{number of ounces in the box}}$$

This procedure is referred to as unit pricing. Many stores have implemented unit pricing, but somehow, with their various specials and changing prices, their efforts never seem to cover all items. And fresh produce and some meats and frozen goods are not unit priced at all. Back to my cereal problem; I want to make the best buy. If I look at the ratio for the unit price of the giant size, I get $91¢/_{13\,oz.}$ (You might want to recall that 1 lb. is equal to 16 oz.) But $91¢/_{13\,oz.}$ corresponds to 7¢ per oz. The regular size has a unit price of $69¢/_{10\,oz.}$, or 6.9¢ per oz. I don't save anything by buying the giant size (although the manufacturers naturally assume I'll just believe it is cheaper and will therefore buy *more* of their product). In MWF, we made up ratio problems that we had actually encountered in the past week and exchanged them with each other to gain practice.

I recall one woman who brought in a similar comparison from a local Washington, D.C. grocery chain. In her example, not only did the supereconomy size turn out to be just as expensive as the smaller size, but the smaller size was significantly *cheaper!* So comparison shopping, using ratios, can save you money and save you from buying more than you need.

Formulating your own ratio or comparison problems can be very helpful. Most in the program found that making up their own problems gave them practice in applying basic skills to actual daily situations.

Sometimes we made up our own problems during class and traded with each other. I recall being asked to do problems ranging from calculating the average number of attorneys married to social workers in the Washington, D.C. area, to comparing the number of trees suffering from Dutch elm disease to the number of healthy trees on Pennsylvania Avenue. The variety of problems and the creativity of the members of the class were always amazing.

Making up your own problems or trading problems with someone else can be very helpful. It can provide you with realistic problems, and solving them even can be fun. As you compare notes, see who gets the best buy, and attempt unusual problems, such as the ones we shared in MWF.

The following cases will give you practice in using ratios to make comparisons.

Determine which of the following is a better buy:
(a) A 28 oz. can of tomatoes for 59¢, or a 16 oz. can for 39¢?
(b) A 13 oz. can of tuna for $1.89, or a 7 oz. can for 99¢?
(c) A 65 oz. box of detergent for $2.45, or a 32 oz. box for $1.79?
(Answers on page 94.)

PROPORTIONS

Ratios are particularly useful in solving many consumer problems and also problems in science and business. Especially useful will be the case where we have two ratios set equal to each other. Whenever we have this equality between two ratios, we say we have a *proportion*.

Example 3: For example, we would say that $\frac{12}{8} = \frac{36}{24}$.
This is true since $12 \cdot 24 = 36 \cdot 8$
$$288 = 288$$
You may want to recall from Chapter 3 that if we have $\frac{a}{b} = \frac{c}{d}$, then we can say that from

$$\frac{a}{b} = \frac{c}{d} \qquad \text{means} \quad ad = bc$$

In our particular example $a = 12, b = 8, c = 36$, and $d = 24$. There are four members in a proportion, and if any three of them are known, it is possible to find the missing number. The proportion $\frac{a}{b} = \frac{c}{d}$ can also be written as: $a : b = c : d$. Sometimes a double colon is used to represent the proportion: $a : b :: c : b$. We usually read such a statement this way: "*a* is to *b* as *c* is to *d*," where:

$$a \longleftrightarrow a$$
"*is to*" $\longleftrightarrow \div$
$$b \longleftrightarrow b$$
"*as*" $\longleftrightarrow =$
$$c \longleftrightarrow c$$
"*is to*" $\longleftrightarrow \div$
$$d \longleftrightarrow d$$

Answers to page 93:

(a) $59¢/28$ oz. $= 2.11¢/$oz. $=$ better buy; $39¢/16$ oz. $= 2.44¢/$oz.

(b) $189¢/13$ oz. $= 14.54¢/$oz.; $96¢/7$ oz. $= 14.14¢/$oz. $=$ better buy

(c) $245¢/65$ oz. $= 3.77¢/$oz. $=$ better buy; $179¢/32$ oz. $= 5.59¢/$oz.

Example 4: Consider the following question: What number is to 5 as 12 is to 30?

Before solving the above word problem we again recall our 5-step guide to solving word problems:

Step 1   READ AND REPRESENT
Step 2   GUESS
Step 3   TRANSLATE
Step 4   SOLVE
Step 5   CHECK AND CONCLUDE

I cannot stress enough that this five-step method should merely act as a guide for developing your own problem-solving method:

Step 1   Read and Represent

The unknown element in our problem is the number, which I have represented by n.

Step 2   Guess

As an estimate we might note that 12 is a little less than half of 30. Therefore, we could possibly estimate n to be a little less than one-half of 5.

Step 3   Translate

Keeping in mind that $a/_b = c/_d$ is read as "a is to b as c is to d."
We get "What number is to 5 as 15 is to 30?"
This becomes $n/_5 = {}^{12}/_{30}$.

Step 4   Solve

We again keep in mind that $a/_b = c/_d$ means $ad = bc$.
We have $n \cdot 30 = 5 \cdot 12$
$$n \cdot 30 = 60$$
$$n = \frac{60}{30} = 2$$

Step 5   Check and Conclude

We know that 2 is certainly a little less than one-half of 5. Our answer is certainly reasonably close to our estimate. If we reduce the fraction $^{12}/_{30}$ we get $^2/_5$. Our conclusion is this:

2 is to 5 as 12 is to 30

The following word problem is a case in which a proportion is used to answer a particular question:

Suppose that in a certain school the ratio of pupils to teachers is 32 to 3, or 32 : 3. If there are 160 pupils present on a given day, how many teachers would you expect to be present?

Step 1   Read and Represent

In reading and rereading our problem we want to represent our unknown element, in this case the number of teachers which we will represent by x.

Step 2   Guess

At this step we might note that if the ratio of pupils to teachers is 32 : 3, or $^{32}/_3$, that ratio is approximately $^{10}/_1$, or 10 to 1; therefore, we could expect there to be 10 times as many pupils as teachers. If we have 160 pupils, we might estimate x to be $^1/_{10}$ of 160, which is 16 teachers.

Step 3   Translate

Recall that in this step we are translating verbal relationships into math relationships. The ratio of pupils to teachers is 32 : 3. This means that the ratio of 160 pupils to x teachers must be equal to 32 : 3. We have a proportion: $160 : x = 32 : 3$. Rewritten in terms of fractions we have:

$$\frac{160}{x} = \frac{32}{3}$$

Step 4   Solve

To solve $^{160}/_x = {}^{32}/_3$ we merely take

$$160 \cdot 3 = 32 \cdot x$$
$$480 = 32x$$
$$15 = \frac{480}{32} = x$$

Step 5   Check and Conclude

To check, we note that $^{160}/_{15} = {}^{32}/_3$ and we can therefore conclude that there are 15 teachers in this school.

A typical proportion problem of great interest to every driver is *gas mileage!* Take, for example, the following:

Example 5: If a car travels 138 miles on 6 gallons of gasoline, how many miles does it travel on 8 gallons?

Step 1   Read and Represent

Let m represent the unknown quantity, which is the number of miles traveled.

Step 2   Guess

Estimating an answer to this question is not as easy as it was in some of our previous problems. However, we might observe that 138 is about 20 times 6. Therefore, m will also be around 20 times 8, or 160.

Step 3   Translate

If we reword our problem in terms of proportions, it can actually be restated as follows:

138 miles is to 6 gallons as m miles is to 8 gallons.
Translating we get:

$$\frac{138}{6} = \frac{m}{8}$$

Helpful hint: In every proportion both ratios must be in the same order. If we take miles to gallons on the left, we must also take miles to gallons on the right.

Step 4   Solve

As we saw in previous problems

$$\frac{138}{6} = \frac{m}{8} \quad \text{means}$$

$$138 \cdot 8 = 6 \cdot m$$

$$1{,}104 = 6m$$

$$184 = \frac{1{,}104}{6} = m$$

Step 5 Check and Conclude

The figure 184 is certainly close to our approximation of 160. You may want to check for yourself that

$$\frac{138}{6} = \frac{184}{8}$$

Our conclusion is that our car will be able to travel 184 miles on 8 gallons of gasoline.

In connection with gas mileage and traveling, reading and understanding maps often involves solving proportions, as in the next example.

Example 6: On a road map of New York State, $\frac{1}{2}$ inch represents 6 miles. How many miles are represented by $1\frac{1}{4}$ inches?

Step 1 Read and Represent

Let m again represent the number of miles involved.

Step 2 Guess

If $\frac{1}{2}$ inch represents 6 miles, we can say that 1 inch will represent 2 times as many miles, or 12 miles. Thus, $1\frac{1}{4}$ inches should be a little over 12 miles.

Step 3 Translate

Rewording our problem we see that $\frac{1}{2}$ inch is to 6 miles as $1\frac{1}{4}$ inches are to m miles. Translating this statement becomes

$$\frac{\frac{1}{2}}{1\frac{1}{4}} = \frac{6}{m}$$

Step 4 Solve

$$\frac{\frac{1}{2}}{1\frac{1}{4}} = \frac{6}{m}$$
$$\frac{1}{2} \cdot m = 1\frac{1}{4} \cdot 6$$
$$\frac{1}{2}m = (\frac{5}{4}) \cdot 6$$
$$\frac{1}{2}m = \frac{30}{4}$$
$$2 \cdot \frac{1}{2}m = 2 \cdot (\frac{30}{4})$$
$$m = \frac{60}{4} = 15$$

Step 5   Check and Conclude

Our conclusion is that 1 ¼ inches on a map of New York represents 15 miles.

CONVERSIONS—THE METRIC SYSTEM

Ratios and proportions are also useful tools in making conversions. Schools are introducing the metric system in the elementary grades, some recipes are being written in terms of liters, and odometers and speedometers in new cars give distance and speed in both miles and kilometers. Table 4.1 gives some of the equivalent expressions in our present system and the metric system. Proportions are going to allow us to make conversions not listed in the table.

Example 7: Suppose you were asked to find the number of grams in .75 ounces.

Step 1   Read and Represent

Let m = number of grams.
When we look at Table 4.2, we see that 1 ounce is equal to 28.3 grams. We might say that m grams is to 1.5 ounces as 28.3 grams is to 1 ounce. (Note that we need to be careful that if we take grams to ounces on the left, we also take grams to ounces on the right.)

Step 2   Guess

Since .75 ounces is ¾ of an ounce, and 1 ounce equals about 28 grams, we might guess that .75 ounces corresponds to ¾ of 28, or $(¾) \cdot 28 = {}^{84}/_4 = 21$ grams.

Step 3   Translate

m grams is to .75 ounces as 28.3 grams is to 1 ounce becomes

$$\frac{m}{.75} = \frac{28.3}{1}$$

Step 4   Solve

To solve the above equation we have

$$m \cdot 1 = (.75) \cdot (28.3)$$
$$m = 21.2 \text{ (to the nearest tenth)}$$

TABLE 5.1

ENGLISH UNITS AND METRIC UNITS		
Length		
1 millimetre (mm)	≈ .04 inches (in)**	or 1in = 25.4 mm
1 centimetre (cm)	≈ .4 inches (in)	or 1 in = 2.54 cm
1 metre (m)	≈ 1.1 yards (yd)	or 1 yd ≈ .92 m
1 kilometre (km)	≈ .62 miles (mi)	or 1 mi ≈ 1.6 km
Mass		
1 gram (g)	≈ .035 ounces (oz)	or 1 oz ≈ 28.3 g
1 kilogram (kg)	≈ 2.2 pounds (lb)	or 1 lb ≈ .45 kg
Volume		
1 millilitre (ml)	≈ .03 fluid ounces (fl oz) or 1 fl oz ≈ 30 ml	
1 litre (l)	≈ 1.06 quarts (qt) or 1 qt ≈ .95 l	
1 cubic metre (m^3)	≈ 1.3 cubic yards (yd^3)or 1 yd^3 ≈ .76 m^3	
Area		
1 sq centimetre (cm^2)	≈ .16 sq inches (in^2) or 1 in^2 ≈ 6.5 cm^2	
1 square metre (m^2)	≈ 1.2 square yards (yd^2) or 1 yd^2 ≈ 0.8 m^2	
1 square kilometre (km^2)	≈ 0.4 square miles (mi^2) or 1 mi^2 ≈ 2.6 km^2	

*We have chosen to use the international spellings for metric system units as recommended in the *Interstate Consortium on Metric Education: Final Report* (California State Department of Education, Sacramento, 1975).

**The symbol ≈ means "is approximately equal to." The information in this table is from the United States Department of Commerce, National Bureau of Standards, Special Publication 330, revised 1974.

COMMON COMPARISONS WITHIN THE METRIC SYSTEM	
Length	
1 millimetre (mm)	= $\frac{1}{1000}$ metre (m)
1 centimetre (cm)	= $\frac{1}{100}$ metre (m)
1 kilometre (km)	= 1000 metres (m)
Mass	
1 milligram (mg)	= $\frac{1}{1000}$ gram (g)
1 kilogram (kg)	= 1000 grams (g)
Volume	
1000 millilitres (ml)	= 1 litre (l)

Step 5   Check and Conclude

We can see at a glance that our approximation was very close. Again, I will leave the check to you to determine that the answer is actually correct.

Our conclusion, therefore, is that .75 ounces is equal to 21.2 grams.

We see that future conversions to the metric system will require facility for working with proportions, even if a table has been provided.

Ratios and proportions allow us to make comparisons between sales items, different systems of measurement, and purchases. You probably cannot get through an entire day without dealing with these comparisons. Just the other day, while I was eating lunch with some of my colleagues the subject of gas mileage came up in the conversation. We were discussing the self-service gas stations. I proudly mentioned that I had finally pumped my own gas. Since I am not very mechanical, the thought of going to a gas station and fumbling around, trying to follow the directions while everyone is in line *waiting* always resulted in anxiety. I could just picture myself spilling gas all over the ground and myself, or not being able to turn the pump on to begin with. Finally, finances forced me to go to a self-service station when the price of regular went over a dollar a gallon; I couldn't put it off any longer. With pride, I was able to find my gas tank, fill it, and replace the pump. I must admit that I did have trouble turning the pump on, but a kind fellow traveler came to my aid.

My fear of going to a self-service gas station is quite similar to the fear of math felt by many adults. A lack of confidence in my own, in this case, mechanical, ability was preventing me from saving money. One step needed to overcome such a fear is to attempt the project rather than ignore it. The satisfaction I felt at pumping my own gas can be compared to the satisfaction you are getting from doing a math problem correctly.

To do something you thought you never could do is immensely rewarding. Your self-image improves, and you may find that this new self-confidence will carry over into other projects.

Getting back to lunch and ratios, there I was at lunch with other members of the college faculty. A friend of mine, Ellen, who holds a Ph.D. in psychology, mentioned that she thought her gas tank held 20 gallons. But when she went to the self-service station, with her gas gauge near the empty mark, she discovered that it took 23 gallons to fill the tank. Now Ellen asked (keep in mind that I was the only mathematician at the table) if that meant she was getting better or worse gas mileage than she had thought, since her tank held 23, not 20, gallons.

If there is any experience I find somewhat traumatic, it is being put on the spot to answer a math question, especially when I am trying to enjoy my lunch. No answer came popping into my head, and anxiety was beginning to build as everyone looked toward me, waiting for the answer from the new mathematician. I didn't have a ready answer, but I suggested we take a simpler example.

Suppose Ellen had calculated that her tank held 1 gallon, and she got 4 miles to that 1 gallon. Suppose she had misjudged the size of her tank and that it actually held 2 gallons. That meant that she was getting 4 miles to 2 gallons or $\frac{4}{2}$ or 2 miles to one gallon. Her mileage was worse, and I was off the hook.

Notice that no one wanted the exact mileage; they just wanted to know if it was better or worse. If I had let my immediate reactions gain control over me, I would have given up and switched the topic of conversation. However, you can see how, by taking a very simple case, restating the problem in your own words, and taking it slowly, step by step, you can sometimes arrive at a logical conclusion. Do not be intimidated by questions from friends or colleagues. If I had not been able to answer the question, I might have felt foolish, but I doubt that anyone else would really have given it a second thought. Don't allow your fears to keep you from making a good attempt, whether shopping with a friend, talking to your accountant, or discussing job-related projects. Don't be afraid to take that first step. Take a simple version of the problem. You may not need the exact numerical answer. Reword the problem in a way that makes sense to you. Take your time. And above all, if you don't come up with an answer, don't worry about it. It may come to you later.

Furthermore, those around you probably do not have an immediate answer either. To overcome math anxiety you are going to have to do math. That initial step is necessary. Each time you attempt a math problem, you will gain practice and gain more confidence. Avoiding situations, such as my lunch problem, will only serve to reinforce your anxieties. In the following section I have listed a variety of comparison problems involving ratios and proportions. If you get stuck, refer to our examples in this chapter.

Some of these are actual problems created by participants in MWF. Remember that ratios are comparisons written in the form of fractions. Proportions are merely two ratios that are equal to each other. After doing some of these comparisons, you may want to formulate comparison problems of your own. Take comparisons from grocery shopping, determine your own gas mileage, use ratios to read distances on a road map, compare sales advertised in the newspaper. The list of possible comparisons is endless.

Comparison problems will get easier and can even be fun as you try to beat the manufacturers at their own game. Realizing that two giant-size packages are cheaper than one super-size not only gives you satisfaction at solving a math problem, but you can also gloat over having outwitted the detergent company!

COMPARISON PROBLEMS

(1) If a woman who earns $8,600 a year spends $2,580 a year on food, what is the ratio of her food expenses to her income?

(2) If 3 cans of beets cost 69¢, how much will 5 cans cost?

(3) If pica type on a typewriter gets 10 characters to an inch, how many characters will there be in 5 inches?

(4) If the ratio of Democrats to Republicans is 5 to 3 in a certain county, how many Republicans are there if the county has 20,000 Democrats?

(5) If a recipe calls for ⅔ cup of flour for a main dish to serve two, how much flour is needed to increase the recipe to serve 7?

(6) If it takes 12 gallons of gasoline to drive a car 216 miles, how many gallons will it take to drive 360 miles at the same speed?

(7) Find the number of yards in 10 meters.

(8) If you are informed that your new coffee maker takes 3 tablespoons of coffee for every 6 oz. of water, how many tablespoons of coffee are needed for 10 oz. of water?

(9) A photograph measuring 8 inches by 10 inches is to be enlarged so that its length is 25 inches. What will be its new width?

(10) If a consumer information director receives 125 information requests in 5 days, how many requests should he or she receive in 20 days?

(11) Hugh is knitting a sweater. He finds that he gets 4 rows to one inch. How many rows will he have if he knits 3 ½ inches of his sweater?

(12) On a street map of Syracuse, New York, ½ inch represents ⅝ of a mile. How many miles are represented by 4 inches?

CHAPTER 6
PERCENTAGES

A new house is selling for $36,000,
but requires a down payment of 20 percent.
What is the actual amount of the down payment?

Percentage problems such as the one above can often be confusing. Like many adults, I was taught to solve such problems by
forming fractions, where one number goes in the numerator and
another goes in the denominator.

As a grade school student, I always found this method somewhat unsatisfactory because I was always forgetting which number was to be divided by which. If problems were changed slightly
(for example, if the down payment of 20 percent is $28,000, what
is the selling price?), I became even more confused, because now
I was given the percentage.

Many adults in the MWF program had had similar experiences. Previous work with percentages had been exercises in
memorization and arithmetic. There was little explanation given
for the reasoning behind the method; it just worked.

Percentages turned out to be one of the most enjoyable and
successful topics in MWF. Instead of relying on memorization we
applied the techniques of translation and problem solving from
previous chapters to percentage problems.

These problems turned out to be quite easy using these

methods, and all in the program were very successful with this topic. The most memorable success story with percentage problems came from a woman on the administrative staff at a local university, whom I will refer to as Helen. Helen was in charge of obtaining grants for an international program at the school. Part of her job entailed determining the subsequent budget for the program. After determining the budget, she then had to submit it to her superior, a rather intimidating gentleman who dealt with all grant proposals.

The day after our percentage session in MWF she had to meet with this man about one of her proposals. When she presented him with her budget, involving thousands of dollars, he replied that a certain percentage was incorrect. Quickly he calculated the amount on his calculator and obtained an answer. Somehow this answer didn't seem quite right to Helen, and she said so. This questioning of his calculation seemed to irritate her superior, who immediately recalculated his figure on the calculator and obtained the same result.

Who could dispute the calculator?

Helen became a little flustered by the almost indignant attitude of her superior and also by the speed with which figures were being tossed around. Yet, confident of her own figures and practice in working with percentages in MWF the night before, she merely replied, pencil and paper in hand, "Let's take x...." And with that start, she finished the problem, showed her own figures to be correct, and saved her program thousands of dollars.

Helen's experience revealed some of the effects that math anxiety can have in daily situations.

First of all, her experience showed how a lack of confidence in your own math ability could lead to intimidation by others in a job situation, *even when you are correct*.

Second, there seems to be a great belief in the infallibility of calculators. Calculators will give you correct answers, *if you push the right buttons*. Somehow individuals come to believe that if the problem is done on a calculator, there is no room for error, forgetting the possibility of their own possible errors in setting up the problems. The math-anxious adult may tend to accept any result obtained with these machines.

Third, the speed with which the director was trying to do the calculation was not only intimidating but could have prevented Helen from getting a chance to think out her own approach. By saying, "Let's take x," she not only took control of the situation away from the director, but was able to proceed at her own speed. Helen's success reinforced our claim that confidence in your own ability can be as important as the actual skills involved.

We have spoken a lot about how much anxiety can prevent you from making job changes and career advancements. Perhaps we have not really dwelt on the fact that math anxiety can contribute to a poor all-around self-image. Math anxiety can allow you to be intimidated by others even when you are right and they are wrong. It takes confidence in yourself and assertiveness to be able to contradict someone else's claims. The other person may not be willing to accept your challenge, and persistence will be needed.

Percentages turned out to be an area where techniques in problem solving and translation could be applied quite easily. Most of the adults in the program found that percentage problems actually were easy! This topic was particularly helpful in building confidence and positive attitudes.

In addition, there was a great deal of interest in doing percentage problems. Banking, business, budgets, mortgages, sales, discounts, loans, and, of course, taxes were just some of the everyday experiences involving percentages. With more and more women working, family budgets and planning for the future become more complicated. Everyone needs to be able to understand and use percentages.

The rest of this chapter will deal with the three basic types of percentage problems that one can encounter. Techniques and methods to solve them will be given in addition to confidence-building tips found successful in MWF. Before attempting applied problems such as discounts, taxes, and interest, we will begin by acquainting ourselves with the terminology involved in percentage problems and simple examples.

So, as Helen said, "Let's take x. . . ."

The word *percent* is derived from two Latin words, *per centum*, and simply means per hundred or hundredths. If a survey reveals that 75 percent of all adults suffer from some form of math anxiety, then we can say that 75 out of 100 adults or $^{75}/_{100}$ adults are subject to that anxiety. Percent usually is denoted by a percentage sign (%). Thus, we have the following equivalent expressions:

$$75 \text{ percent} = 75\% = \frac{75}{100}$$

Note that a percent can always be rewritten as a fraction whose denominator is equal to 100:

Thus, 5 percent = 5% = $\frac{5}{100}$ = $\frac{1}{20}$.

Here are some examples:

$$10\% = \frac{10}{100} = \frac{1}{10}$$
$$20\% = \frac{20}{100} = \frac{2}{10} = \frac{1}{5}$$
$$25\% = \frac{25}{100} = \frac{1}{4}$$
$$50\% = \frac{50}{100} = \frac{1}{2}$$
$$75\% = \frac{75}{100} = \frac{3}{4}$$
$$90\% = \frac{90}{100} = \frac{9}{10}$$

It is also often convenient to be able to rewrite percents as fractions and vice versa. For example, suppose we want to rewrite $\frac{3}{4}$ as a percent. We want to answer the question:

$\frac{3}{4}$ is what percent? The unknown in this question is the percent, let x = percent. Using our translation techniques from Chapter 2 we have the following:

$\frac{3}{4}$ is what percent
becomes $\frac{3}{4}$ = x%
becomes $\frac{3}{4}$ = $x/_{100}$

Now we have a simple proportion to solve:

$$\frac{3}{4} = \frac{x}{100}$$

$$4x = 300$$

$$x = \frac{300}{4}$$

$$x = 75$$

Our answer is that $\frac{3}{4}$ is equivalent to 75 percent or 75%.

We also note that whole numbers, which can be considered fractions whose denominators are equal to 1, can also be rewritten as percents. For example, 3 is what percent? Again, letting x stand for the percent, we have

$$\frac{3}{1} = {}^{x}/_{100} \quad \text{Solving for } x, \text{ we get}$$
$$300 = x$$
$$\text{or 3 is } 300\%$$

Sometimes, such as with interest on savings accounts, we encounter percentages involving fractions. For example, at one bank a depositor receives $5\frac{1}{4}$ interest. In order to calculate how much interest he or she receives on the balance, it is necessary to first rewrite $5\frac{1}{4}\%$ as a fraction. We want to answer the question, $5\frac{1}{4}\%$ is what fraction? Let x be this number. Translating, we get

$$x = \frac{5\frac{1}{4}}{100} = 5\frac{1}{4} \div 100$$
$$= \frac{21}{4} \div 100$$
$$= \frac{21}{4} \cdot \frac{1}{100}$$
$$= \frac{21}{400}$$

The table below gives some of the fractional equivalents of percentages.[1]

TABLE 6.1		
$10\% = {}^{1}/_{10}$	$12\frac{1}{2}\% = {}^{1}/_{8}$	$8\frac{1}{3}\% = {}^{1}/_{12}$
$20\% = {}^{1}/_{5}$		
$40\% = {}^{2}/_{5}$	$25\% = {}^{1}/_{4}$	$16\frac{2}{3}\% = {}^{1}/_{6}$
$50\% = {}^{1}/_{2}$	$37\frac{1}{2}\% = {}^{3}/_{8}$	$33\frac{1}{3}\% = {}^{1}/_{3}$
$60\% = {}^{3}/_{5}$	$62\frac{1}{2}\% = {}^{5}/_{8}$	$66\frac{2}{3}\% = {}^{2}/_{3}$
$75\% = {}^{3}/_{4}$	$87\frac{1}{2}\% = {}^{7}/_{8}$	$83\frac{1}{3}\% = {}^{5}/_{6}$

You may want to practice changing the following percents to fractions:

$$(1) \quad \tfrac{1}{2}\% = {}^{1}/_{200}$$
$$(2) \quad 3\tfrac{1}{3}\% = {}^{1}/_{30}$$
$$(3) \quad 4\% = {}^{1}/_{25}$$
$$(4) \quad 7\tfrac{1}{2}\% = {}^{3}/_{40}$$
$$(5) \quad 1\tfrac{1}{2}\% = {}^{3}/_{200}$$

Change fractions to percents:

(1) $^{64}/_{100}$ = 64%
(2) $^5/_{50}$ = 10%
(3) $^5/_8$ = 62.5%
(4) 2 = 200%
(5) $^1/_6$ = 16.7%

Traditional presentations of percentage problems have always involved memorizing an equation: $P = R \cdot B$ and solving for one of the unknown terms. We are going to dispense with this approach and rely solely on our translation techniques from Chapter 2, plus our previous experiences with percentages in banking, business, sales, taxes, and many other everyday practices.

It turns out that there are really three basic types of percentage problems.

THREE BASIC TYPES OF PERCENTAGE PROBLEMS

Finding a Percent of a Given Number

What is 5% of 120?

We begin this problem by recalling the suggested steps in solving word problems from Chapter 3:

Step 1 READ (the problem) AND REPRESENT
(unknown quantities)
Step 2 GUESS
Step 3 TRANSLATE
Step 4 SOLVE
Step 5 CHECK AND CONCLUDE

Step 1   READ AND REPRESENT

Rereading this problem, we see that "what" actually refers to "what number." Our unknown is a number, which I have chosen to represent by n.

Step 2   GUESS

Recalling our procedures in Chapter 3, in this step we try to make an intelligent guess. We might note that 5% of 100 is 5. Therefore, since 120 is slightly larger than 100, we might assume that 5% of 120 will be slightly larger than 5. Again the use of the term *slightly* certainly is relative, and our conclusion is not an accurate one. However, such guessing or analyzing has two major benefits. First of all, it gives you control over the problem. You have obtained a common sense, approximate answer. Sometimes it is not even necessary to have the exact answer. If the 5% represents interest saved in a bank account for Christmas shopping, knowing that you will have a little more than $5 will be sufficient. Second, your approximation gives you a means of checking your answer and prevents you from making unnoticed large arithmetic errors. For example, if you end up with an answer of $60, you will know immediately that your answer doesn't make sense.

This guessing step is not as vague or hit-or-miss as it seems. What you are really doing in this step is using your intuition or previous experience to make an intelligent approximation.

Step 3   TRANSLATE

Translating, "What is 5% of 120" becomes

$n = {}^5\!/_{100} \cdot 120$ **where**

n corresponds to ⟶ what number

= corresponds to ⟶ is

${}^5\!/_{100}$ corresponds to ⟶ 5%

of corresponds to ⟶ $\cdot$ (multiplication)

Step 4   SOLVE

Determining n can be done by solving this equation:

$$n = {}^5\!/_{100} \cdot 120$$

$$n = \frac{5 \cdot 120}{100}$$

$$n = \frac{600}{100}$$

$$n = 6$$

Step 5   CHECK AND CONCLUDE

If possible, it's always helpful to check your answer, either by an alternate method or by comparing your answer with your estimate in Step 2. The answer 6 certainly seems reasonable with our guess of a number slightly larger than 5.

Our conclusion is that 6 is 5% of 120. The following problems are three examples of this type of percentage problem:

(1) What is 3% of $400?

(2) What is 15% of 40?

(3) What is 25% of 80?

Answers are given below.

Finding What Percent One Number Is Of Another

56 is what percent of 80?

The above problem represents the second category in basic percentage problems. Again we apply our problem solving steps to find the desired answer.

Step 1   READ AND REPRESENT

Upon rereading the problem we note that "what percent" means "what number percent." Again, I chose n to represent the desired number.

Step 2   GUESS

Guesses at possible answers will, of course, vary from individual to individual. The purpose of this step is to give a reasonable estimate of the solution based on past experience, intuition, or common sense. In this particular problem you might note that 40 is 50 percent or one-half of 80. Therefore, 56 will be somewhere between 50 percent and 100 percent. This certainly does not give us an exact answer, but it does put the solution within a certain range of numbers.

$$\text{ANSWERS:}$$
$$(1)\ m = \frac{3}{100} \cdot \$400 = \frac{1200}{100} = 12$$
$$(2)\ \frac{15}{100} \cdot 40 = \frac{600}{100} = 6$$
$$(3)\ \frac{25}{100} \cdot 80 = \frac{2000}{100} = 20$$

Step 3   TRANSLATE

"56 is what number percent of 80"
becomes

$$56 \; = \; \frac{n}{100} \cdot 80$$

Step 4   SOLVE

Solving the above equation we get:

$$56 = \frac{n80}{100}$$
$$5600 = n \cdot 80$$
$$70 = n$$

Step 5   CHECK AND CONCLUDE

We conclude that 56 is 70 percent of 80. We can check this answer by taking 70 percent of 80 which is

$$\frac{70}{100} \cdot 80 \text{ or } \frac{5600}{100} \text{ or } 56.$$

The following problems are characteristic of this second basic percentage problem. Again answers are shown below.

Problems: (1)   12 is what percent of 60?

              (2)   50 is what percent of 20?

              (3)   What percent of 24 is 12

Answers:

(1) $12 = \dfrac{n}{100} \cdot 60$

      $60 \cdot n = 1{,}200$

      $20 = \dfrac{1{,}200}{60} = n$

(2) $50 = \dfrac{n}{100} \cdot 20$

      $5{,}000 = n \cdot 20$

      $250 = \dfrac{5{,}000}{20} = n$

(3) Note that the order of the words has been changed. However, the translation gives us the same basic type of percentage problem.

"What percentage of 24 is 12?" becomes

$$\frac{n}{100} \cdot 24 = 12$$
$$n \cdot 24 = 1{,}200$$
$$n = \frac{1{,}200}{24} = 50$$

Finding A Number When A Percent Of It Is Known

60 is 30 percent of what number?

Third in our list of basic percentage problems, the above example again turns out to be fairly easily solved.

Step 1   READ AND REPRESENT

Our unknown element in this problem is a number which I have represented, as usual, by n.

Step 2   GUESS

Making a logical guess is slightly more difficult with this type of percentage problem. However, we could make the following observations: 30 percent is almost one third. Thus "60 is 30 percent of what number" can be thought of as "60 is $\frac{1}{3}$ of what number." That number would be 3 times 60, or 180.

Step 3   TRANSLATE

This type of percentage problem also translates quite easily:

60 is 30% of what number

becomes

$$60 = \frac{30 \cdot n}{100}$$

Step 4   SOLVE

We use methods from previous chapters to obtain a solution:

$$60 = \frac{30 \cdot n}{100}$$
$$6{,}000 = 30 \cdot n$$
$$\frac{6{,}000}{30} = n$$
$$200 = n$$

Step 5   CHECK AND CONCLUDE

First we check our answer against our intuitive guess, and we can determine that $n = 200$ certainly seems reasonably close to our guess of 180.

We can also check that 30 percent of 200 is equal to $^{30}/_{100} \cdot 200 = {}^{6000}/_{100} = 60$.

Our conclusion is:

60 is 30% of 200

The problems that follow will give you practice in doing percentage problems in this third category. Again, answers are shown at the bottom of the page:

(1) 90 is 45% of what number?
(2) 16 is 20% of what number?
(3) 30% of what number is 18?

The next step in working with percentages is to apply these three basic techniques to everyday problems. As we mentioned earlier there is a wealth of daily activities that require some skill in percentages. Examples ranging from figuring out tips and calculating discounts to determining taxes and interest rates are easy to find.

Answers:

$$(1) \quad 90 = 45 \cdot \frac{n}{100}$$
$$9{,}000 = 45n$$
$$200 = \frac{9{,}000}{45} = n$$

$$(2) \quad 16 = 20 \cdot \frac{n}{100}$$
$$1{,}600 = 20n$$
$$80 = \frac{1{,}600}{20} = n$$

(3) We note that again the order of the words in the problem has been changed. However, the translation step reveals that we still have a percentage problem in the third category:

$$30 \cdot \frac{n}{100} = 18$$
$$30n = 1{,}800$$
$$n = \frac{1{,}800}{30} = 60$$

In working with percentages you probably will find it helpful to recall past experiences and use common sense where helpful.[2]

EVERYDAY PROBLEMS

Let's go back to our original example in the beginning of this chapter:

> A new house is selling for $36,000, but requires a down payment of 20 percent. What is the actual amount of the down payment?

Purchasing homes is becoming more and more difficult because of rising interest rates. Banks are requiring a greater percentage of the total cost as a down payment. As a result more and more families are finding it difficult to meet these rates. Determining just what amount of money is required becomes vital. In doing everyday percent problems we will apply the same tools we developed for answering word problems in general throughout the book.

Step 1   READ AND REPRESENT

In Step 1 we want to read the problem thoroughly to determine what is actually being asked. We may want to rewrite the main question in our own words.

Our goal in this initial step is to determine which of the three basic categories of percent problems our example fits into. We note that the down payment is found by taking the given percent of the total cost. The question "What is the cost of the down payment?" could actually be rewritten as "20 percent of $36,000 is what number?" Our example fits into the first category.

Our unknown is the down payment, which I have chosen to represent by D. Let D = down payment.

Step 2   GUESS

Recalling that percent is really a fraction and that 20 percent corresponds to $20/100$ or $1/5$ we might guess that 36,000 is close to 40,000 and $1/5$ of 40,000 is 8,000. This estimate gives us a good idea of what our answer should be.

Step 3   TRANSLATE

At this point we are again applying the techniques of Chapter 2 to our rewritten question from Step 1.
Thus,

"20 percent of $36,000 is the down payment"

becomes

$$\frac{20}{100} \cdot \$36,000 = D$$

Step 4 SOLVE

Solving the above equation we get

$$\frac{20 \cdot 36,000}{100} = D$$

$$\frac{720,000}{100} = D$$

$$7,200 = D$$

Step 5   CHECK AND CONCLUDE

In this step you want to check your answer against your estimate in Step 2. Our estimate was $8,000, so our answer seems to be a logical one.

Conclusion: Many times we do math problems and stop after having arrived at a numerical solution. In word problems it's nec- essary to interpret your answer in terms of the problem.

This example is fairly straightforward, but we would want to mention that 7,200 stands for $7,200 needed for the down payment on the house.

The next example is one familiar to every adult who has ever eaten in a restaurant.

Tipping

If dinner for 4 at a Chinese restaurant costs $28.40 what tip should the party leave? Let us assume that the tip is 15 percent of the total bill.

Step 1   READ AND REPRESENT

Reading the problem we see that it also involves a type 1 percent problem. We must determine 15 percent of 28.40. I chose T to denote the amount of the tip.

Step 2   GUESS

At this point many find determining 10 percent of 28.40 to be quite easy. Recalling again that 10 percent is really $^{10}/_{100}$ = $^{1}/_{10}$, we see that $^{1}/_{10}$ of $28.40 would be equal to $2.84. So our tip will be slightly larger than $2.84.

Step 3   TRANSLATE

T is 15% of 28.40
becomes
$$T = {}^{15}/_{100} \cdot 28.40$$

Step 4   SOLVE

Now we need only solve the above equation:

$$T = \frac{15 \cdot 28.40}{100}$$
$$T = \frac{\$426.00}{100}$$
$$T = \$4.26$$

Step 5   CHECK AND CONCLUDE

Compared with our estimate, $4.26 does not seem to be an outrageous answer. Therefore we might conclude that the tip should amount to $4.26.

If you and your friends are feeling generous you will probably leave $4.50. No one expects you to leave exactly 15 percent of the bill; an approximate sum will suffice. You might extend this idea of approximation to more areas of your math work. I feel that so much emphasis in math instruction has been placed on getting the exact correct answer that many adults fail to apply math effectively. They are so worried about getting the answer right down to the last digit that any slight errors or complications often result in frustration and failure to complete the problem. In many situations, such as tipping, approximate answers are good enough. If you want a quick estimate of 15 percent of any item, just consult a waiter or waitress, they can give you an answer in a minute. This ability does not mean that all waiters and waitresses are geniuses in mental arithmetic. However, it does indicate that they have devised a method to determine that 15 percent, and they have had lots of practice.

It took a waitress, not another math major, to show me how to calculate 15 percent quickly. First, we can approximate the bill to be around $30. Then we figure that 10% of 30 is $^{10}/_{100} \cdot 30 = \$3.00$. Now, 5% is really ½ of 10%. So 5% is ½ of $3.00 or $1.50. The total tip becomes $3.00 plus $1.50 or $4.50 which is what we left anyway. Do not hesitate to use such schemes in any math problems. Logical shortcuts such as this one will allow you to spend less time with specific calculations and will actually give you better control over the problem. If your scheme makes good sense to you, solving problems will become more of an exercise in reasoning and less of an opportunity for confusion.

Cut-rate drug stores, 50 percent-off sales, 10 percent rebates, and ⅓-off reductions are all examples of one of the major ploys of businesses to take our last dollar—the *discount.*

Discounts

A discount is merely a reduction of the original selling price of an article. The amount taken off is often denoted by percentages. Stores often advertise "Special Sale—33⅓% off."

Let's look at the following case:

> Toasty Stove Company is selling $800 wood-burning stoves at a 40% discount during the summer months. Determined to conserve fuel, you decide to purchase a stove during June. How much will you have to pay?

Step 1   READ AND REPRESENT

The expression 40 percent off implies that the stove company will take 40 percent off the price of the stove before selling it to you. This 40 percent of the $800 original price is your discount. The solution to the problem is going to involve determining the discount and the final net price of the stove. From previous experience we know that this net price will be equal to the original price minus the discount.

Let D represent the amount of the discount. The final net price is also unknown and could be represented by N.

Step 2   GUESS

We might note in Table 5.1 that 40% is about ½. So 40 percent off 800 should be a little less than ½ of 800, which is $400.

Step 3   TRANSLATE

"The discount, D, is 40 percent off of 800
becomes $D = {}^{40}\!/_{100} \cdot 800$.

"The final net price is the original price minus the discount" becomes $N = \$800 - D$

Step 4   SOLVE

We can see that the value of D is needed in order to solve the second equation. So we begin by solving the D-equation first:

$$D = \frac{40 \cdot 800}{100}$$

$$D = \frac{32,000}{100}$$

$$D = 320$$

$$N = \$800 - D$$
$$= \$800 - \$320$$
$$= \$480$$

There is an alternative method for calculating discounts. Let us take our example of the $800 wood-burning stove at 40 percent off. If 40 percent of $800 is deducted you may conclude that 60 percent of $800 is left to pay. Your net price is really 60 percent of 800, or

$$^{60}\!/_{100} \cdot 800 = {}^{48,000}\!/_{100} = \$480.$$

For many adults, in MWF this second and shorter method was more desirable. Still, many chose to get the discount and then subtract the result from the original price.

Again, we want to stress that you must choose methods that work well for you. If your method takes longer but leaves less room for error or confusion, stick with it! Remember that speed is not a major goal in doing math. Understanding and reaching reasonable solutions are the goals.

Percent of Increase or Decrease

Gloria wants to purchase a $695 stereo. Since she does not have that much money available, Gloria plans to pay for the stereo on the installment plan. The stereo dealer says that she can buy the stereo for $148 down and $49 a month for 14 months. What is the amount of additional money, or increase, that Gloria will pay by using the installment plan? What percent of the original price is this increase?

Step 1   READ AND REPRESENT

We can observe that Gloria will need to calculate three items in order to answer the above question. First, she needs to determine the final cost of the stereo, then the amount of increase, and finally what percentage this increase is of the original price.

Let C denote the final cost and I the amount of the increase and X the percent of increase.

Step 2   GUESS

We might note that $49.00 for 14 months is approximately $50 a month or $700. The cost to Gloria, therefore, will be almost $150 plus $700, or $850. This is more than $100 extra that Gloria will be paying for the privilege of buying the stereo on time.

Step 3   TRANSLATE

In this particular translation we have two verbal relationships to translate into mathematical equations. First, we must determine the amount of increase, I, which is the difference between the final cost, C, and the original price. Thus, $I = C - 695.

The second relationship is to determine the percent of increase. It can be rephrased to fit one of our three basic types of percentage problems. To find the percent of increase, we are actually determining what percent the increase is of the original price. Since we are allowing X to denote this percent, we now have:

I is equal to what percent of the original price? Translating, we get

$$I = \frac{X}{100} \cdot 695$$

Step 4   SOLVE

To solve $I = C - 695$ we must first determine C, the final cost of the stereo. C will be obtained by multiplying $49.00 times 14. ($49 for each month) and adding the result to $148. We have:

$$C = 148 + (49)(14)$$
$$= 148 + 686$$
$$= 834$$

Therefore, the increase $I = 834 - 695 = 139$
To find the percent of increase we solve:

$$I = \frac{X}{100} \cdot 695$$

But $I = 139$, so we have

$$139 = \frac{X}{100} \cdot 695$$
$$139 = \frac{X \cdot 695}{100}$$
$$13{,}900 = X \cdot 695$$
$$20 = \frac{13{,}900}{695} = X$$

Our percent of increase is 20 percent.

Step 5   CHECK AND CONCLUDE

Comparing our final answer with our estimate, we see that our original estimate was indeed very close to the actual amount. Our conclusion is actually threefold. The final cost of the stereo is $834, the amount of increase is $139, and this increase is 20 percent of the original price of $695.

Taxes

We have saved the most natural application of percentages for last—taxes. For most of the adults in the MWF program, taxes and percentages go hand in hand. For most, these two words hardly spelled *relief*, but spelled *anxiety*, with a capital *A*. For many, doing taxes is dreaded as much as taking a calculus course. Tables, charts, and percentages, bills, receipts, and forms add up to enough confusion to last you the whole year. No wonder anxiety mounts. The trauma of doing your taxes is increased by any anxiety you have about doing plain percentage problems.

We can not eliminate taxes, but we can at least cut down the anxiety associated with the math involved. You may find that your newly organized approach to percentage problems in general may give you a more organized approach to taxes. With less anxiety and a greater sense of control over the math involved, you will feel more prepared (*armed* might be a better word) to meet the battle of the 10-40. The following example might be one that you could be faced with in April:

Last year Harry had 18 percent of his salary withheld for taxes. If the total amount withheld was $3,168.00, what is Harry's yearly salary?

Step 1   READ AND REPRESENT

We are to find Harry's yearly salary. Rewording the question we have, "$3,168.00 is 18 percent of what salary?"

Our unknown quantity is of course Harry's salary, which could be represented by S.

Step 2   GUESS

Guessing an answer to this question may not be quite as straightforward as with previous examples. We may be able to get a rough estimate, though, by making the following observation. Suppose $3,168.00 was 10 percent of Harry's salary. That would mean that $3,168.00 was $\frac{1}{10}$ of the salary. Therefore, the salary would be 10 times $3,168.00 or $31,680.00. Since $3,168.00 represents 18 percent of S or an even larger part of S, we would expect S to be smaller than $31,680. Again, any common sense estimate you make can be helpful. This step will at least prevent you from making large errors and getting an answer like 12 million!

Step 3   TRANSLATE

"$3,168.00 is 18 percent of what salary?"
becomes

$$3,168.00 = \frac{18}{100} \cdot S$$

Step 4   SOLVE

Solving the equation from Step 3 we get:

$$3,168.00 = \frac{18 \cdot S}{100}$$
$$316,800 = 18S$$
$$\frac{316,800}{18} = S$$
$$\$17,600 = S$$

Step 5   CHECK AND CONCLUDE

We might check our answer by taking

18 percent of 17,600:
$$\frac{18}{100} \cdot 17,600 = \frac{18 \cdot 17,600}{100} = \frac{316,800}{100} = \$3,168.00$$

Our answer checks, and the conclusion would be that Harry's salary must have been $17,600.

You will probably want to practice working on the percent problems that follow.

Again, we want to stress that understanding math problems and their solutions is only one step toward becoming capable and confident. Math does require practice. As one friend of mine said, "You don't *study* math. You *do* math!" Trying to learn percent problems without doing more and more problems on your own would be the same as trying to learn to play the piano by reading the music and watching someone else play. The idea that you either know the math involved or you don't, so why bother doing problems, is just another myth associated with math. Practice doing more and more problems not only helps you develop your problem-solving methods but it also reinforces your confidence. Each successfully completed problem adds to a positive attitude and gives you a sense of accomplishment.

I recall a math teacher I had in high school who said that there was no greater satisfaction or more rewarding feeling than solving a math problem! I must agree. You can write what you feel is a good report or good paper, but you probably won't regard it as perfect. Also, what you feel is a fantastic report may be just mediocre to another. And always it can be refined. However, if you finish a math problem and get the right answer, you know you've got it, 100 percent. (No pun intended here!) To increase confidence and develop math methods—practice!

In the chapter on ratios we mentioned that all adults in MWF spent a good deal of their time making up word problems of their own, exchanging problems, solving them.

Percentage is another area where this technique is very helpful. The newspaper, TV news shows, magazine articles, sales promotion; the list could go on and on. All of these items are good sources for forming problems of your own. You may ask members of your family or friends to suggest percent problems, preferably ones they have actually run across. In formulating your own problems the key to obtaining the correct answer will be determining what is being asked in the initial step. As always, I want to encourage you to follow a method that makes sense to you. Our five-step procedure is merely a guide.

Again, we want to stress that, in working on math problems from any topic, you should begin with simple problems (such as 10 percent of 80 is what number?) and then move on to more difficult ones. Developing math abilities is a progression. You apply techniques for simple problems to more difficult ones and keep on building abilities. You will find that percentages are fairly easy to work with. Being able to use and calculate percentages is an ability needed by more and more adults. Archie Bunker's statement to little Stephanie on "All in the Family" that "Little girls don't need arithmetic" could not be further from the truth. Women and men must rely on math to meet financial obligations and plan for the future. Reducing math anxiety and building a positive approach to everyday math problems can help all of us meet rising inflation, more complicated tax problems, and ever tightening budgets with more confidence.

The following practice problems are ones that were found to be useful in the course.

MORE PRACTICAL PROBLEMS

(1) If a savings account increases from $2,400 to $3,000 over a period of a year, what is the percent of increase?

(2) If a clock radio costing $28 is reduced 25 percent, what is its sale price?

(3) A utility company offers a 4 percent discount on your bill if it is paid within 30 days. How much will you save if you pay a $26 bill within the 30-day period?

(4) Mr. Smith's income tax is 12 percent of his income. Last year he paid $1,920 in income taxes. What was his income for the year?

(5) A shopper noticed that the price of a loaf of bread rose from 85¢ to 90¢ in one week. What is the approximate percent of increase in the cost of the bread?

MORE PERCENTAGE PROBLEMS

(1) What is 2 percent of $150?

(2) What is 35 percent of $500?

(3) What is 75 percent of 120?

(4) Nine is what percent of 36?

(5) Twelve is what percent of 80?

(6) What percent of 40 is 24?

(7) Fifteen is 60 percent of what number?

(8) Eight is 40 percent of what number?

(9) Thirty-six is 200 percent of what number?

(10) A savings bank gives 5½ percent interest on its savings accounts. If you have $600 in your account, what will your interest be?

(11) A man received $120 on an investment of $800. What rate (percent) did he receive?

(Answers to these problems can be found in Appendix B.)

REFERENCES

1. Nielsen, Kaj L. *Mathematics for Practical Use.* New York: Barnes & Noble, 1962.
2. Selby, Peter H. *Practical Algebra.* New York: John Wiley & Sons, 1974.

CHAPTER 7
ALGEBRA REVISITED

In MWF I often asked adults to try to pinpoint the time when their troubles with math began. The most common response I received was *algebra*. Somehow algebra seemed to be the point at which math made simple became math made difficult.

Where arithmetic had consisted of definite rules and concrete examples, algebra seemed to be made up of complicated expressions and abstract ideas. To do well in arithmetic you had to memorize multiplication tables and do countless drill problems: adding fractions, working with decimals, and doing long division. All of a sudden in algebra you had to work with abstract ideas such as exponents, factors, and quadratic equations. Instead of adding and subtracting numbers, you were required to manipulate formulas and expressions similar to $3x^2y$ and $x^2 + 2x + 1$. Memorization and drill were suddenly not enough.

Algebra also required making generalizations and applying new concepts to a variety of word problems. Algebra represented, for many, their first serious encounter with everyday word problems.

As I have stated before, many of the difficulties with these word problems are due to the fact that few explanations or procedures were given in traditional algebra courses on how to apply newly learned techniques and skills to verbal problems. Often too little time was spent on approaching and solving word problems.

Finally, many practical problems assume a background of day-to-day experiences that a high school student has not had time to acquire.

For most adults math is a useful and desirable skill that can be applied to practical situations. Using math in this way means forming and solving word problems.

Only in the classroom do you encounter problems set up in neat series of equations to solve or fractions to add and multiply.

If you view math as only this series of calculations and formulas, you will be at a great disadvantage when you try to use the math you've learned. Suddenly, you can't seem to make the connection between the formulas and operations you memorized and the verbal problem you don't even know how to set up.

No wonder many feel that math can be irrelevant. They aren't able to use the math they've studied. For math-anxious adults, reintroducing themselves to math will require a different approach from the memorization techniques of their school years.

In the Math Without Fear program I found that the key to understanding and applying math skills was the word problem. My whole approach has been to study word problems, how to approach and how to solve them. Then, in the process of attempting various word problems, I've introduced specific math tools: translation techniques, solving basic equations, and working with ratios and percentages.

Actually, without my mentioning it explicitly, you have already devoted a considerable amount of time to algebra. As I stated in Chapter 3, the first time I used the letter x to denote an unknown quantity, I was using algebra. Algebra is simply an extension of arithmetic. In arithmetic you add and multiply numbers, while in algebra you do these same operations except with both letters and numbers. Algebra allows you to solve those same everyday problems I've been talking so much about. You know that $2 + 3 = 5$ from your study of arithmetic. But without algebra how can you answer questions such as this:

If a man sold 3 shares more than $\frac{1}{2}$ of his stock and has 10 shares left, how many shares did he start with?

Obviously, knowing how to add and multiply fractions is not

going to be sufficient for answering this kind of question. We need only to extend these basic math skills by introducing a letter to denote the original number of shares and then applying our five-step guide from Chapter 4.

You may want to take time now to review your new problem-solving skills by doing this problem. I've listed a possible five-step solution below. Remember that your procedure and mine may not be exactly the same. But you're striving to find a method that works best for you.

Step 1   Read and Represent

I chose s to denote the original number of shares.
I also noted that there were three verbal relationships involved.
The amount he sold was ½ of his shares plus 3.
The amount he had left was 10.
The third relationship was implied: that the amount sold plus the amount he had left was equal to the original amount of shares, namely s.

Step 2   Guess

This is a good example of a problem in which a logical guess, or estimate, doesn't necessarily come to mind. In this case I chose to eliminate this step and move on to the next. Notice that I don't allow myself to become anxious because I don't have an estimate. Be flexible. Some steps are extremely useful for particular problems and of little use for others.

Step 3   Translate

In this step I've summarized the three translations needed:

$$\text{Amount sold} = (½ \cdot s) + 3$$
$$\text{Amount left} = 10$$
$$s = \text{amount sold} + \text{amount left}$$
and becomes
$$s = \frac{1}{2} s + 3 + 10$$

Step 4   Solve

$$s = \frac{1}{2} s + 3 + 10$$

Now I refer to my basic equation-solving techniques from Chapter 3.

$$s = \frac{1}{2} s + 3 + 10$$

$$s = \frac{1}{2} s + 13$$

$$s - \frac{1}{2} s = \frac{1}{2} s - \frac{1}{2} s + 13$$

$$s - \frac{1}{2} s = 0 + 13$$

$$\frac{1}{2} s = 13$$

$$2 \cdot \frac{1}{2} s = 2 \cdot 13$$

$$s = 26$$

Step 5   Check and Conclude

Checking we see that $(\frac{1}{2} \cdot 26) + 3 + 10 =$
$13 + 3 + 10 = 26$.
We conclude that the man began with 26 shares.

In this problem and many of the other problems in previous chapters you've been using some of the fundamentals of algebra, including solving some basic equations and working with expressions using both letters and numbers.

In the rest of this chapter you'll be looking more closely at basic algebraic tools and applying them to a variety of word problems needing algebraic techniques not yet covered.

The next word problem requires some new algebraic techniques:

> A homeowner is planning to enlarge the recreation room in her home. The room now measures 3 meters by 4 meters. She wants to enlarge the room by adding the same length to each side. How many meters should she add to each side if the new room is to have an area of 30 square meters?

This is a good example of a problem for which a diagram is really helpful. Originally the room looked like this figure:

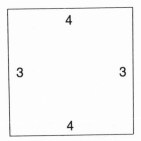

The new room can be visualized in the following diagram:

$x + 4$

$x + 3$ $x + 3$

$x + 4$

It's important to note that the length and width are being measured in meters. You may want to recall that a meter is a unit of length in the metric system, approximately three inches longer than a yard. (Refer to Table 5.1 in Chapter 5.)

Step 1   Read and Represent

The unknown, the amount to be added to the length and width of room, has already been represented by x. The only information we have to work with is the area of the room.

You probably remember working with area problems in the past. The area of a rectangle, whether of a floor or box or garden, is found by multiplying the length times the width.

The area is always stated in square feet, square inches, or, as in this case, square meters. In Chapter 8, I will be considering areas in more detail.

Coming back to our example, I see that the area of the new room is 30 square meters. What are the dimensions of the new room? In other words, what are its length and width? The length of the room is $x + 4$, and the width is $x + 3$. The area, 30 square meters, equals the length times the width, or $(x + 4)$ times $(x + 3)$.

Step 2   Guess

For the moment I'll leave out this step and go on to Step 3.

Step 3   Translate

Translating I get $30 = (x + 4)(x + 3)$.

At this point I note that doing the rest of the problem will require solving the above equation. Furthermore, it appears that my basic techniques of equation solving are not going to be sufficient for solving this new type of equation. You may recall that to solve an equation you derive a series of simpler equivalent equations until you finally reach $x = \square$ or $\square = x$.

In this particular case I note that I first need to explain what $(x + 4)(x + 3)$ really means. It turns out that $30 = (x + 4)(x + 3)$ is equivalent to $30 = x^2 + 7x + 12$. Getting from this equation to $x = \square$ or $\square = x$ is not at all obvious and will require the introduction of new math vocabulary and more basic tools of algebra. I actually began this step in Chapter 3. You probably recall that a letter used to denote a number is a *variable*; a *term* was either a single number or the product or quotient of one or more variables and/or numbers. For example, $2y$, $5xz$, and 11 are all terms.

ALGEBRAIC EXPRESSIONS

An *algebraic expression* is a collection of terms connected by a plus or minus sign. For example, $2y + 5xz + 11$ would be an algebraic expression with three terms: $2y$, $5xz$, and 11. An *equation* is, therefore, a mathematical statement, or sentence, that two algebraic expressions are equal.

In our previous example $30 = x^2 + 7x + 12$ is a math sentence stating that the algebraic expressions 30 and $x^2 + 7x + 12$ are equal. Solving the equation means finding the values of x, which make the math sentence a true statement. Our goal will be to find a procedure to solve $30 = x^2 + 7x + 12$. To do this you will need to know what x^2 stands for; you will need more practice in adding, subtracting, and multiplying various algebraic expressions; and finally you must know how to solve equations with not just x or y terms, but x^2 and y^2 terms also.

Let us begin by introducing some new words into our math vocabulary. Suppose we have two terms: $3x$ and $5x$. The numbers 3 and 5 in these terms are referred to as the *numerical coefficients* of x. As I mentioned earlier, algebra is really just a generalization of arithmetic, using the same operations of addition, subtraction, multiplication, and division, but working with both numbers and letters.

It's conceivable that in working with algebra I may want to multiply two terms. For example, I may want to multiply $3x$ by $5x$. I want to simplify $3x \cdot 5x$. Recall that $3x$ is an abbreviation of "3 times x," or $3 \cdot x$. The same is true of $5x$. My expression is really $3 \cdot x \cdot 5 \cdot x$. Can it be simplified? First, I note that the order in which I multiply numbers does not matter (refer to Appendix A). Take, for example, $(3 \cdot 2) \cdot 5 = 3 \cdot (2 \cdot 5)$, in which we do the multiplication inside the parentheses first. Both sides are equal to 30. Thus, $3 \cdot x \cdot 5 \cdot x$ is the same as $3 \cdot 5 \cdot x \cdot x$. But $3 \cdot 5$ is 15, and we have $15 \cdot x \cdot x$. To simplify this expression, mathematicians introduced a new notation, referred to as an *exponent*. They abbreviated $x \cdot x$ as x^2, where 2 is called the *exponent* of x, x is called the *base* and x^2 is read as x squared. This idea can be extended to multiplying x by itself any number of times: $x \cdot x \cdot x \cdot x \cdot x \cdot x = x^6$ and is read as x to the sixth power. To multiply our two terms $3x$ and $5x$, we end up with the single term $15x^2$.

Suppose you were asked to multiply $2xy$ by $5yz$. This product could be expressed as $2 \cdot x \cdot y \cdot 5 \cdot y \cdot z$. Rearranging, we would obtain $2 \cdot 5 \cdot x \cdot y \cdot y \cdot z$. But $2 \cdot 5$ is actually 10, and $y \cdot y$ is y^2. Our result is $10 \cdot x \cdot y^2 \cdot z$, or $10xy^2z$. Multiplying two terms together results in another term, and we write this term as a product with a single numerical coefficient.

What about adding terms such as $3xy$, $2z$, and $5xy$? This sum could be expressed as $3xy + 2z + 5xy$. Can it be simplified? In Chapter 3, I spoke briefly about *like terms*. For example, in the expression $2 + 4x + 7 + 5x$, 2 and 7 are like terms. They each consist of a single number. Similarly, $4x$ and $5x$ are like terms because they are both of this form: a number times x. We can define like terms in general: *like terms* are terms that are exactly the same except for their numerical coefficients. Thus, $3xy$ and $5xy$ would be like terms. However, $2z$ and $5xy$ are *unlike terms*. In adding and subtracting algebraic terms, only like terms can be combined. In $3xy + 2z + 5xy$ the first and last terms can be combined, and our expression becomes $8xy + 3z$. You may want to think of xy in $3xy$ and $5xy$ as some object. In MWF I often compared adding $3xy$ and $5xy$ with adding 3 objects plus 5 more of the same object. For example, xy might stand for an apple. If I asked you what 3 apples plus 5 apples would give you, you naturally would answer 8 apples. If you let xy stand for an object of some kind, 3 of these objects plus 5 more will represent 8 of these objects. Therefore, $3xy + 5xy$ is $8xy$.

Let's look at another example. Suppose we have $5b^2 - 2b^2 + 3ab$. Can this algebraic expression be simplified? To answer this question you will have to determine if there are any like terms. You'll notice that $5b^2$ and $2b^2$ are like terms and that 5 objects minus 2 of the same object gives 3 objects or $5b^2 - 2b^2 = 3b^2$. Hence, $5b^2 - 2b^2 + 3ab = 3b^2 + 3ab$. This expression cannot be simplified further since $3b^2$ and $3ab$ are not like terms.

As one final example for this section let us take this algebraic expression: $2x^3 + 5x^2 - 3xy + y^2 - 3$. This expression consists of five terms. At first glance my reaction was that such a long expression could surely be simplified. However, as I looked more closely I noticed that there were no like terms! Therefore, it could not be simplified further. Below you will find a variety of algebraic expressions to be simplified. Answers are given below.

(1) $2x^2 + 3x^2 - x^2 + x$
(2) $5a^2b + 3ab^2$
(3) $2y^4 + 3y^3 + 3y^4 + 2x^3$
(4) $8b^6 + 6b^5 + 3b^6$
(5) $x^2 - 2x + 3x - 1$
(6) $2x + 3y - 2x - 6y + 3$
(7) $3x^2 + 8y^2 - x^2 - y^2$
(8) $3ab^2c^3 + 2a^2b^3c - ab^2c^2 + a^2b^2c^3$

Answers: (1) $4x^2 + x$ (2) Already simplified (3) $5y^4 + 3y^3 + 2x^3$ (4) $11b^6 + 6b^5$ (5) $x^2 + x - 1$ (6) $-3y + 3$ (7) $2x^2 + 7y^2$ (8) already simplified

At this point I think it's important to mention that I will not be reviewing all of algebra in this single chapter. The purpose of this chapter is to reacquaint you with more of the basic symbols, words, and operations used in algebra. In keeping with the rest of *Math Without Fear* this chapter is supposed to make you feel comfortable with algebraic expressions. And, of course, one of the major goals is to approach algebra with the idea of using it to solve everyday problems.

After you have completed this chapter you will have gained the confidence and some of the basic tools needed to do a more thorough treatment of algebra with a traditional textbook or in a classroom situation. You will be able to approach algebra with a better understanding of what algebra is, how and why it works, and how to apply it.

Studying mathematics can be viewed as a building-block process. You begin with some definitions and symbols. Then you derive new expressions by combining two or more of your definitions; you continue the process of deriving new terms and combining these expressions to arrive at new ones. Each step is built upon material introduced in a previous step.

So far in your reintroduction to algebra you've seen how to multiply two terms together. Another common name for a term is *monomial,* or an algebraic expression with one term.

The second operation you have revisited is the process of combining two or more terms and simplifying the result.

For example, $2x^2y + 5x^2y + 6$ is equal to $7x^2y + 6$.

This final algebraic expression also has a name in the language of math. It is referred to as a *polynomial.* A polynomial is an algebraic expression with two or more terms, where each term is the product of letters and/or numbers.

The next logical operation to consider would be to determine the product of not just two monomials, but of a monomial and a polynomial. Let's look at the following example:

Suppose that in doing a word problem, a particular translation results in this expression: $3x (x + 5)$ Can this expression be simplified? The answer is yes.

We refer to Appendix A and recall a property of real numbers. If I am given a product such as $2 (3 + 5)$, this product can be determined by removing the parentheses and multiplying each number by 2 and simplifying. Thus, 2 (3 + 5) becomes $2 \cdot 3 + 2 \cdot 5 = 6 + 10 = 16$. This property is referred to as the *distributive property*. It also can be used with algebraic expressions. The only difference between arithmetic expressions and algebraic ones is that letters are used in algebra. Since these letters stand for real numbers, you can apply the same rules and properties that work for numbers. For $3x$ $(x + 5)$ you can remove the parentheses and multiply each term in $x + 5$ by $3x$. Then $3x$ $(x + 5)$ becomes $3x \cdot x + 3x \cdot 5$. The term xx can be denoted by x^2. Since it is customary to write algebraic expressions with the numerical coefficient first, you can make use of the fact that $3x \cdot 5 = 5 \cdot 3x$. Note that $5 \cdot 3x = (5 \cdot 3) \cdot x = 15x$. Finally, we can represent $3x$ $(x + 5)$ by $3x^2 + 15x$.

Helpful hint: To know which operations to do first in an arithmetic expression such as $(2 \cdot 5) + (3 \cdot 7)$, you must first consider the quantities inside parentheses as single numbers. To simplify the expression, do the work within the parentheses *first*, and then add the new terms.

The   expression $(2 \cdot 5) + (3 \cdot 7)$ becomes $(10) + (21) = 31$. What if there had been no parentheses? Would $2 \cdot 5 + 3 \cdot 7$ still be equal to 31? Someone might be tempted to think of $2 \cdot 5 + 3 \cdot 7$ as 2 times (5 + 3) times 7, which would be 2 times 8 times 7, or 112.

To eliminate such confusion, mathematicians adopted the convention of doing all multiplication and division operations *first* from left to right, and then doing addition and subtraction. That way $2 \cdot 5 + 3 \cdot 7$ could be simplified *only* to 10 + 21, or 31. Going back to the previous example, $3x \cdot x + 3x \cdot 5$ can be simplified only as $3x^2 + 15x$.

Before moving on to the next section, try some of the problems given below. (Answers on page 137.)

MULTIPLYING A POLYNOMIAL BY A MONOMIAL
Simplify

    (1) $2 (5 + x)$
    (2) $(x + y) 3$
    (3) $3 (2xy + y + 6)$
    (4) $(6 + x) 2x$
    (5) $x (x + 5 + 3x)$
    (6) $(2x + 5y + 6) 6y$

One more word about notation. It's customary to write a polynomial involving only one letter with the term with the largest exponent of x first, and then the remaining terms with decreasing exponents and any numerical terms last. Thus, $4 + 2x^2 + 3x$ is usually written as $2x^2 + 3x + 4$. This does not mean that one version is more correct than the other. They are equally correct, but you'll find it's often useful to simplify further problems by first writing the expression in this fashion. Once again, this step of rearranging the order of $3 + 2x^2 + 5x$ is perfectly acceptable, since the order in which you add real numbers can also be rearranged.

Continuing the building-block process, the next step will involve multiplying a binomial by another binomial. A *binomial* is a polynomial consisting of exactly two terms.

MULTIPLYING A BINOMIAL BY ANOTHER BINOMIAL

Suppose you want to simplify this expression:

$$(x + 4)(x + 3)$$

There are two different methods you can use to simplify the above expression.

In the first method, place one of the two expressions below the other:

$$x + 3$$
$$\underline{x + 4}$$

To find the product, multiply each term in $x + 3$ by each term in $x + 4$. This step gives you two different expressions:

4 times $x + 3$ is equal to $4x + 12$
x times $x + 3$ is equal to $x^2 + 3x$

Answers to page 136:
(1) $10 + 2x$
(2) $x \cdot 3 + y \cdot 3 = 3x + 3y$
(3) $6xy + 3y + 18$
(4) $12x + 2x^2$
(5) $x^2 + 5x + 3x^2 = 4x^2 + 5x$
(6) $12xy + 30y^2 + 36y$

These two expressions, or products, are given below with like terms placed under each other as indicated. Next, these two expressions are added to obtain the final answer:

$$
\begin{array}{r}
x + 3 \\
x + 4 \\
\hline
4x + 12 \\
x^2 + 3x \\
\hline
\hline
x^2 + 7x + 12
\end{array}
$$

You may want to note the similarity of this procedure to the one used in arithmetic problems such as the multiplication problem below:

$$
\begin{array}{r}
15 \\
\times \quad 12 \\
\hline
30 \\
15 \\
\hline
180
\end{array}
$$

In our second method we find the product $(x + 4)(x + 3)$ by applying the distributive property twice: $(x + 4)(x + 3)$ is $(x + 4)$ times x plus $(x + 4)$ times 3. Applying the distributive property again to each of these two products, we get $x^2 + 4x$ plus $3x + 12$ or $x^2 + 4x + 3x + 12$. Adding like terms, our final answer is $x^2 + 7x + 12$. In this method it is important to make sure that each term in $(x + 3)$ is multiplied by each term in $(x + 4)$. This means that there are four different products to be calculated and then added. As an aid to prevent you from missing one of these products you can use a popular method known as the FOIL method. In the FOIL method you use each letter in the word FOIL to indicate one of the four products that have to be calculated.

In the example $(x + 4)(x + 3)$
F indicates multiplying *FIRST* terms: $x \cdot x = x^2$.
O indicates multiplying *OUTERMOST* terms: $x \cdot 3 = 3x$.
I indicates multiplying *INNERMOST* terms: $4 \cdot x = 4x$.
L indicates multiplying *LAST* terms: $4 \cdot 3 = 12$.

Then these products are added to obtain $x^2 + 7x + 12$. This method applies *only* to the special case of multiplying two binomials. Let us take another example and form the product by both the first method and by the FOIL method:

$$(x + 1)(x + 8)$$

First Method

$$x + 8$$
$$\underline{x + 1}$$
$$x + 8$$
$$\underline{x^2 + 8x}$$
$$x^2 + 9x + 8$$

FOIL Method

$$(x + 1)(x + 8)$$
$$F = x \cdot x = x^2$$
$$O = x \cdot 8 = 8x$$
$$I = 1 \cdot x = x$$
$$L = 1 \cdot 8 = 8$$

Adding: $x^2 + 8x + x + 8 =$
$$x^2 + 9x + 8$$

Listed below are some products you can quiz yourself on.
(1) $(a + 3)(a + 1)$
(2) $(x + 5)(x + 3)$
(3) $(x + 1)(x + 6)$
(4) $(2x + 1)(x + 3)$
(5) $(6 + b)(5 + b)$
(6) $(a + 2)(xa + 7)$
(7) $(t - 2)(t + 4)$
(8) $(a + b)(a + b)$

In problem 8, you have probably discovered that
$$(a + b)(a + b) = a^2 + ab + ab + b^2$$
$$= a^2 + 2ab + b^2$$

(7) $t^2 + 2t - 8$ (8) $a^2 + 2ab + b^2$
(5) $30 + 11b + b^2$ (6) $a^2 + 9a + 14$
(3) $x^2 + 7x + 6$ (2) $x^2 + 8x + 15$ (4) $2x^2 + 7x + 3$ (1) $a^2 + 4a + 3$

Answers:

You can also show this with pictures! When possible it's nice to be able to see how two expressions are equivalent. Regard the square below:

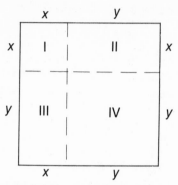

The area of this square is $(x + y)(x + y)$.

This square can be divided into four parts:

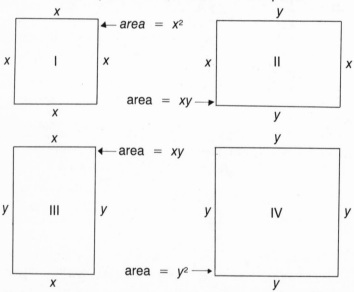

Because these four parts make up the whole square, we may write:

$$(x + y)(x + y) = x^2 + xy + xy + y^2 = x^2 + 2x + y + y^2$$

It's important to keep in mind that you are developing these skills with the goal of applying them to word problems. You'll probably want to refer back to the homeowner's example. In the translation, one arrives at an equation of the form $30 = (x + 4)(x + 3)$, which you now know is equivalent to $30 = x^2 + 7x + 12$. The rest of this section is devoted to more techniques for solving this type of equation.

In the previous section, I dealt with finding the product of two binomials, such as $(x + 1)(x + 2)$ to get $x^2 + 3x + 2$. Note that, after simplifying, the result always has three terms. It is called a *trinomial*. Now I want to consider the reverse procedure. Suppose I have a trinomial to start with, for example, $x^2 + 10x + 16$. Can I write it as the product of two binomials? Can I write $x^2 + 10x + 16$ as $(x + \Box)(x + \Box)$, where each $\Box$ stands for a number? You probably recall this process as *factoring*. That is, factoring is the process of writing an algebraic expression as a product of terms. Each term is called a *factor*.

Let us suppose that $x^2 + 10x + 16$ factors as
$$(x + c)(x + d).$$
Using the FOIL method, we see that this product is equal to $x^2 + dx + cx + cd$. Adding like terms, we get $x^2 + (c + d)x + cd$, which is supposed to equal $x^2 + 10x + 16$. In order to have this equality we see that $(c + d)$ will have to equal 10, and cd will have to equal 16. Factoring $x^2 + 10x + 16$ becomes a matter of finding the appropriate c and d.

To organize your search for c and d, first list all possible c and d that will give you $cd = 16$. There are 3 possibilities:

$$cd = 16$$
$$c = 1 \text{ and } d = 16$$
$$c = 2 \text{ and } d = 8$$
$$c = 4 \text{ and } d = 4$$

The next step requires looking at the possibilities for c and d in the first step and determining which possibility satisfies our other requirement, namely that $c + d = 10$:

$$c = 1 \text{ and } d = 16 \text{——} c + d = 17$$
$$c = 2 \text{ and } d = 8 \text{——} c + d = 10$$
$$c = 4 \text{ and } d = 4 \text{——} c + d = 8$$

From the above cases we see that only the case in which $c = 2$ and $d = 8$ satisfies $c + d = 10$. Therefore, $x^2 + 10x + 16 = (x + 2)(x + 8)$, and we say that these two terms are the factors of our given trinomial.

Summarizing the details of the previous section, we might say that to factor a trinomial such as $x^2 + 10x + 16$ we do the following steps:

(1) Find all possible c and d so that $cd = 16$.

(2) Test all of the possibilities in step (1) to find c and d so that $c + d = 10$.

(3) $x^2 + 10x + 16$ can now be factored as $(x + c)(x + d)$.

A word of caution: I mentioned earlier that the purpose of this chapter is to reacquaint you with *some* of the basic tools of algebra. This factoring procedure applies only to trinomials in which the numerical coefficient of x^2 is 1. For cases such as $2x^2 + 9x + 4$ you will want to refer to an algebra textbook. I will say more about reading math books later.

The next example will involve factors in which one of the numbers c and d is positive and one is negative (see Appendix A for material on negative numbers).

Suppose we want to factor this expression:

$$x^2 + x - 6$$

Factoring this expression means that we must first find c and d so that $cd = -6$. You will note from Appendix A that c and d must have different signs. For this example there are four possibilities:

$$cd = -6$$
$$c = -1 \text{ and } d = 6$$
$$c = 1 \text{ and } d = -6$$
$$c = -2 \text{ and } d = 3$$
$$c = 2 \text{ and } d = -3$$

Next we want $c + d$ to be equal to 1, and we test each of our four possibilities for c and d. (Note that the numerical coefficient of x^2 is 1.)

$$c = -1 \text{ and } d = 6 \qquad c + d = 5$$
$$c = 1 \text{ and } d = -6 \qquad c + d = -5$$
$$c = -2 \text{ and } d = 3 \qquad c + d = 1$$
$$c = 2 \text{ and } d = -3 \qquad c + d = -1$$

Only the case in which $c = -2$ and $d = 3$ satisifies our requirements. We can then conclude that

$$x^2 + x - 6 = (x + (-2))(x + 3) = (x - 2)(x + 3)$$

Note that $x + (-2)$ is equivalent to $x - 2$ (see Appendix A).
Listed below are trinomials that can be factored. Answers follow.

(1) $x^2 + 3x + 2$
(2) $a^2 + 5a + 6$
(3) $a^2 + 12a + 35$
(4) $t^2 + 4t - 12$
(5) $y^2 - 2y - 15$

SOLVING EQUATIONS USING FACTORING

Returning once again to the homeowner example, it will be necessary to solve this equation:

$$30 = x^2 + 7x + 12$$

In Chapter 3 you solved equations in one letter where the letter was always x. Now you have an equation with an x^2 term. Notice that you can apply procedures of Chapter 3 to get zero on one side of the equation and all other terms on the opposite side. This is done by subtracting 30 from each side: $30 = x^2 + 7x + 12$ becomes $0 = 30 - 30 = x^2 + 7x + 12 - 30$, which can be written as $x^2 + 7x - 18 = 0$.
Any equation of the form $ax^2 + bx + c = 0$ is called a *quadratic equation*. For this example $a = 1, b = 7,$ and $c = -18$.

Before we continue with solving the given equation, it will be necessary to state one of the really interesting and useful properties of real numbers. This property deals with zero. If two numbers are multiplied together and equal zero, one or both of them must equal zero. (That is, if $cd = 0$, either $c = 0$ or $d = 0$). For example, you could have $0 \cdot d = 0$ or $c \cdot 0 = 0$. How can we use this fact in solving $x^2 + 7x - 18 = 0$?

First, we note that $x^2 + 7x - 18$ can be factored into $(x + 9)(x - 2)$ using the earlier methods from this chapter. But $x^2 + 7x - 18 = 0$. So, $(x + 9)(x - 2) = 0$. If I consider $(x + 9)$ as c and $(x - 2)$ as d, I know that either $c = 0$ or $d = 0$, that is, $x + 9 = 0$ or $x - 2 = 0$.

Now I have reduced $x^2 + 7x - 18 = 0$ to two separate equations. If we solve either of these equations, I will have a solution to the original equation. Solving I get this:

$$
\begin{array}{c|c}
x + 9 = 0 & x - 2 = 0 \\
\text{becomes } x + 9 - 9 = 0 - 9 & \text{becomes } x - 2 + 2 = 0 + 5 \\
\text{or} & \text{or} \\
x = -9 & x = 2
\end{array}
$$

Step 4 Solve

In the original homeowner problem solving has produced two possible answers: $x = -9$ or $x = 2$.

Step 5 Check and Conclude

But what did x represent in the word problem? If you recall, x denoted the length to be added to each side of the house. A length of -9 does not make any sense. We discard that solution. It is precisely in these kinds of problems that this step can be so vital. Checking the other solution of $x = 2$ you can determine that the new addition will be $4 + 2$ by $3 + 2$, or 6 by 5 meters. Its area, length times width, will be $6 \times 5 = 30$ square meters, which is correct. Almost one chapter later you can conclude that the dimensions of the new addition will be 5 meters by 6 meters!

To give you some practice in solving basic quadratic equations, more examples are listed below. The answers follow.

(1) $x^2 + 10x + 21 = 0$
(2) $x^2 + 9x + 8 = 0$
(3) $0 = b^2 + 2b - 8$
(4) $y^2 + 6y + 8 = 0$
(5) $x^2 + 3x - 10 = 0$

Answers: (1) $(x + 3)(x + 7) = 0$, so $x = -3$, or $x = -5$
(2) $(x + 1)(x + 8) = 0$, so $x = -1$, or -7
(3) $(b + 4)(b - 2) = 0$, so $b = -4$, or -8
(4) $(y + 2)(y + 4) = 0$, so $y = -2$, or $b = 2$
(5) $(x - 2)(x + 5) = 0$, so $x = 2$, or $x = -4$

Helpful hint: Before giving another example of a word problem utilizing the new equation-solving tools of this chapter I want to say a few words about reading mathematics books. After progressing through *Math Without Fear* you will have the confidence and skills needed to learn more math and to apply it. Perhaps you will even enjoy using various math skills and you may want to go further. For example, you may decide that you want to do more algebra. Math books can be intimidating. And even though you've gained confidence in your own abilities, anxieties can creep in when you begin to read a math book, especially if you begin to read it like any other book.

You cannot skim a math book. It must be read slowly and with paper and pencil at hand. Choose a book to meet your level of math preparedness. If you want to review fractions, you obviously do not want to use a book that assumes you know algebra. After the first pages in your chosen text, you will begin to meet concepts and ideas you don't understand. Don't let anxieties set in! After all, if you already understood these ideas, you wouldn't need the book! Instead, reread the material, try a few examples, sit back and think about it! Don't worry if progress seems slow at this point; other pages will go faster. However, you should not be tempted to jump to other easier sections. Mathematics builds upon a succession of steps, each step using the results of previous ones. Math books build, also. Tools to solve problems in Chapter 5 will come from Chapters 1 to 4. Skipping around to different parts of the book will be confusing, and old anxieties are sure to return. Finally, do not restrict yourself to one source. You can use two or more texts to help you develop a particular area of math. Different explanations may be helpful.

Many everyday problems require solving quadratic equations. The next example is one such problem.

Example: The Statewide Security Company manufactures burglar alarms for the home. Their profit is made by selling b, burglar alarms, and is given by this formula: $b^2 + 6b - 16$. How many alarms must the company sell in order to break even?

Step 1   Read and Represent

Fortunately, the unknowns and verbal relationships have already been represented. All we have to do is interpret the phrase, "break even." From past experience you know that breaking even is the same as having a profit of zero dollars. To answer the question we need to solve this equation: $0 = b^2 + 6b - 16$.

Step 2   Guess

I might note that if I sell one detector, the profit is $(1)^2 + 6 (1) - 16$, or $1 + 6 - 16$, or $7 - 16$, or -9 dollars. The company loses money if only one detector is sold.

   If 10 detectors are sold (i.e., $b = 10$), then the profit is $(10)^2 + 6 (10) - 16$, or $100 + 60 - 16$, or $144. After comparing the profits in these two cases, I might guess that the number of detectors that the company must sell is between one and ten.

Step 3   Translate

Great! All the work is already done.

Step 4   Solve

$$0 = b^2 + 6b - 16$$
$$0 = (b + 8)(b - 2)$$

$b + 8 = 0$ | $b - 2 = 0$
$b + 8 - 8 = 0 - 8$ | $b - 2 + 2 = 0 + 2$
$b = -8$ | $b = 2$

Step 5   Check and Conclude

We reject the answer $b = -8$, since selling a negative number does not make sense. Thus, I leave it to you to check that by selling 2 detectors, the Statewide Security Company breaks even.

   The problems below are practical ones that can be solved using tools from this chapter. Be sure to remember the five-step guide:

Step 1   READ AND REPRESENT
Step 2   GUESS
Step 3   TRANSLATE
Step 4   SOLVE
Step 5   CHECK AND CONCLUDE

EVERYDAY PROBLEMS:

(1) A landscaping company is to design a formal garden having an area of 80 square feet. The garden is to have a length that is 2 feet greater than its width. What must be the dimensions of the garden?

(2) In a woodcraft shop the total cost in dollars for producing wooden picture frames is given in this equation:

$$x^2 + 3x + 10$$

where x represents the number of frames made. If the shop can spend $50 on this project, how many frames can be produced?

(3) A travel agent is planning to send a local tour group to Bermuda. The profit to be made by the travel agent for this trip can be represented by $x^2 + 2x - 120$. The letter x stands for the number of people going on the trip. How many people are needed for the travel agent to break even (that is, how many people are needed for profits, $x^2 + 2x - 120$, to equal 0)?

CHAPTER 8

WHAT'S THE ANGLE? GEOMETRY AND MEASUREMENTS FROM A DIFFERENT POINT OF VIEW

Suppose that the dining area of your apartment is 6 feet wide and 10 feet long. You want to buy wall-to-wall carpeting for this area. How much will it cost you if the carpeting is $12 per square foot?

From previous experience both with examples from this book and with day-to-day activities as a consumer, homeowner, or apartment dweller, you have probably run across at least one problem similar to the one above. Maybe it was buying a rug, tiling the bathroom floor, putting in insulation, or planting a vegetable garden. Also, you probably realize that solving such problems entails working with dimensions and areas of some geometric figure. Probably no area of math has had so many practical uses as geometry!

Geometry, which is simply a study of shapes and their relationships, was one of the first aspects of math to be developed. The word geometry came from the Greek and literally means "to measure the earth." Most of the written information about geometry comes from the Egyptians and Babylonians. Egyptians used geometry to survey land for tax purposes, to build pyramids, and to carry out various other activities such as making sundials. Today geometry is used in many everyday situations and many careers. Naturally, surveying and architecture rely heavily on geometry. But geometry also has many applications to all persons who use measurements, such as plumbers, electricians, navigators, cooks, florists, fashion designers, and car manufacturers. The many uses of geometry are all around us.

However, the applications of geometry are certainly not restricted to measuring. Geometry also has many aesthetic and psychological effects. The shape of a structure contributes to much more than just the serviceability of the building, appliance, or tool. The shape also contributes to its visual appeal. For example, an American car obviously is not designed solely for its mechanical purposes. Its visual appeal to the buyer is of extreme importance. Architecture is another area where the visual effects of a structure may be just as important as the sturdiness, durability, and other practical aspects. Sometimes it seems that buildings are being designed for their negative visual appeal. I've always thought that the countless new square and rectangular glass and concrete buildings of Washington, D.C., and probably every other large city, seemed to display a lack of creativity or inspiration. According to Hackworth and Howland, it's an accepted fact that boxlike figures in general, arranged in precise regular patterns, often are associated with feelings of boredom or loneliness. This brings up another effect of geometry in everyday experience: its psychological effect on our feelings and emotions.

The shape of an object can greatly determine our emotional reactions to it. Jagged lines, $\wedge\wedge\wedge\wedge\wedge$, usually create feelings of excitement and activity, whereas curves, $\zeta\diagdown\mathcal{O}$, are usually associated with feelings of relaxation and pleasant moods or emotions.

One area that has always relied on the visual and emotional effects of geometric shapes is art. Geometry is rooted in basic ideas such as symmetry and proportion. Hackworth and Howland note that these concepts are particularly important to the artist. They help the artist answer such questions as these: Is a square or a rectangle better suited for a painting? Should the focus be in the center or off center? If off center, how far off center? Many other artists, in addition to painters, must concern themselves with shapes and patterns. Craftspeople such as weavers, potters, persons who do crocheting, needlepoint or macrame are all especially reliant on patterns, symmetries, and proportions. Geometry has been extremely important in contributing to the usefulness, beauty, and psychological effects of all kinds of objects and structures.[1]

After sharing ideas and thoughts with a large variety of adults in the Math Without Fear program, I came to believe that geometry is the area of math that adults have enjoyed the most or have been most successful with in the past. Furthermore, I think the reasons for this "unusual" enjoyment of math are precisely those visual and practical effects of geometry.

When you talk about the basic concepts in geometry—for example, points, lines, rectangles, and triangles—you can actually draw a picture or diagram. I can certainly visualize a triangle and its uses better than $2x^2 + x + 1$! Geometry seems to make sense. I can observe geometric concepts all around me, even in nature. The geometric design of a beehive, or the symmetry of the shape of a butterfly 🦋 or a starfish display geometric patterns. Geometry seems to be one area of math where all of us can apply our own observations and past experiences with various geometric concepts. We can picture these concepts, and, most importantly, we can see a variety of uses for these ideas in everyday situations.

This chapter should be one that you can be relaxed with and can easily begin to enjoy. New math tools and ideas will not be abstract or vague. You will be able to connect these concepts with various daily activities. And finally, you will be able to apply these results to a variety of everyday problems. Geometry is an excellent area for you to continue to build self-confidence and a more relaxed attitude toward math in general.

I begin this chapter by applying the techniques of previous chapters. To study geometry I naturally have to acquaint or reacquaint myself with its basic terms. I want to continue my development of the language of math by learning the language of geometry. In this chapter we will discover how to recognize geometric patterns from previous experiences and how to use these patterns in everyday applications.

THE LANGUAGE OF GEOMETRY

The most basic of all geometric shapes is the point. In fact, the point is so basic that one cannot define it! Instead, I have to be content with having an intuitive idea of what it is. A point is a position in space that I can represent by a dot $(\cdot)$. The point is the most basic geometric shape because all other geometric figures can be considered a collection of points. For example, a *line* ⟵———⟶, is a collection of points that has no width, no thickness, and infinite length. The symbol ◄────►denotes the line; the arrows show that the line extends indefinitely in both directions.

There are a few other terms you will need to build your geometric vocabulary. The first term is *line segment*. If you have the line (l), A B, with the points A and B on l, the collection of all points between A and B is referred to as the *line segment* with *endpoints* A and B. It is often denoted by $\overline{AB}$. If I consider the line (l), with only the point A, ◄──•──►, I can see that A divides the line into three parts: the half-line to the left of A, the point A itself, and the half-line to the right of A. A *ray* is made up of a point A on a line along with either of the half-lines beginning at A. You might think of a ray as a giant searchlight beam. The light is A, and together with the beam it represents a ray. To denote a ray, we usually pick another point along the half-line, say C, and let AC denote the ray represented in the diagram below. Notice that the bar above $\overline{AB}$ denotes a line segment, while an arrow above $\overrightarrow{AC}$ denotes a ray extending indefinitely in one direction.

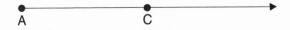

A   C

In addition to the *point* and the *line,* there is another very basic geometric figure you will want to be familiar with: the *plane.* A plane is a flat surface. One usually pictures a plane as the top of a table or desk that extends lengthwise and sideways without end. It has length and width, but no depth.

Still another term basic to the study of geometry is *angle.* An *angle* is formed by the union of two rays and a common endpoint. It can be pictured by the diagram below:

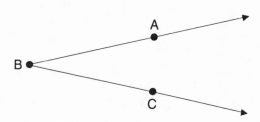

Many common geometric figures can be defined in terms of points, line segments, and planes.

To derive a list of common geometric figures that often occur in practical problems, it will be necessary to define a particular class of figures of which the triangle will be an example. A *plane curve* is a set of points in a plane that can be drawn without lifting your pencil from the paper. All the curves considered in this chapter will be plane curves. I will just refer to curves, meaning plane curves. A curve is *simple* if it can be drawn without retracing any of its points, with the exception that the endpoints may coincide. Figures *a* and *b* below are *simple plane curves,* while *c* and *d* are not:

A *simple closed curve* is one that can be drawn or traced so that you begin and end at the same point. In the diagram above, drawing *b* is a simple closed curve and *a* is not closed. These rather unusual definitions will allow you to define the collection of geometric figures known as polygons.

A *polygon* is a simple closed plane curve that is formed by three or more line segments with common endpoints. Almost every adult has had contact with a variety of polygons, including *triangles,* polygons with three sides; *quadrilaterals,* polygons having four sides, of which rectangles and squares are two examples; and *pentagons*, polygons having five sides (and also seen near Washington).

As mentioned earlier, one of the major practical uses of geometry is its application to measurements. To find a measurement a number is associated with a geometric figure in some way. The basis for measurements will be the mathematical concept of *congruence.* Two geometric figures are *congruent* if they have the same size and shape; in other words, if one figure is an exact copy of the other. For example, the two triangles below are congruent, but the two squares are not. The two squares have the same shape, but definitely are not the same size. You use the idea of congruence in many everyday activities, even if you don't refer to it as such. When you want to replace a worn item, whether it be a part of a car or a television, a battery, or a lightbulb, you need to get an item with the exact size and shape of the old one. The idea of two objects being "alike" is familiar to any adult.

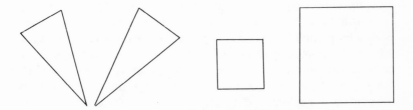

USING GEOMETRY FOR MEASUREMENTS

By now you are probably wondering when these definitions will stop! Well, the end is near. All of these new terms are needed to speak the language of geometry. The concept of congruence will be needed to develop a theory for measurement. Suppose I wanted to measure the length of the line segment below, $\overline{AB}$. To indicate its length, I will need to compare it with the length of a given segment, $\overset{\bullet}{C} \qquad \overset{\bullet}{D}$ (call it $\overline{CD}$). Line segment $\overline{CD}$ can be placed over line segment $\overline{AB}$. Now $\overline{AB}$ is made up of three parts, and each of these three parts is congruent to $\overline{CD}$. Note that each has the same length as $\overline{CD}$.

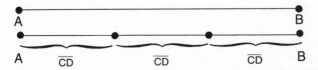

I can express the length of $\overline{AB}$ as "three of the lengths of $\overline{CD}$." For example, if the length of $\overline{CD}$ was an inch, I could express the length of $\overline{AB}$ as "3 inches." A measurement has two parts: a measure, which is a number, and a unit of measure. For example, in the expression "A woman is 65 inches tall," 65 is the measure and the inch is the unit of measure. The measure of a line segment will be found by comparing it with a unit that is also a line segment. The number associated with the length of any line segment will depend upon the size of the unit measure used.

If two persons want to compare the measurement of distances such as the length of a room, the size of a dress, or the height of a table, they must agree on a standard unit of measure.

To meet this need the International Metric System was introduced. (I already referred to this system in Chapter 5. You may want to refer once again to Table 5.1 in that chapter.)

The English system of measurement used in the United States today has many disadvantages. Working with the various units is often very involved. There are 5,280 feet in one mile, 4 quarts in a gallon, and 16 ounces in a pint, to name but a few. In the metric system you can convert from one unit to another by multiplying the given unit by powers of ten. In Chapter 7, you noted that the symbol x^2 could be read as x to the second power and means the product $x \cdot x$. In general, x^n can be read as x to the nth power and stands for the product $\underbrace{x \cdot x \cdot \ldots \ldots \ldots x.}_{n \text{ times}}$

The basic unit of length is the meter. It originally was defined as a line segment one ten-millionth of the distance from the North Pole to the equator. Now it is defined as 1,650,763.73 wavelengths of the orange-red line of the spectrum of krypton -86. You will probably want to think of the meter as being slightly longer than one yard.

Helpful hints:
1. 10 inches is about 25 centimeters
2. A gallon of gasoline is a little less than 4 liters.
3. A quart of milk is a little less than a liter.
4. A nickel weighs about 5 grams and has a diameter of about 2 centimeters.
5. A pound of butter weighs a little less than ½ kilogram.
6. A 110-pound girl weighs about 50 kilograms.
7. A kilometer is a little more than one-half mile.

PERIMETERS

Now that you have considered the measurement of the length of a line segment, you can begin to extend the idea of length to geometric figures. The *perimeter* of any geometric figure is the distance around it. In particular, the perimeter of a polygon, in which all of the sides are straight, is the sum of the lengths of the sides. Let us consider the perimeter of some specific polygons.

In the case of the square, all sides are the same length, L. Thus, its perimeter is $L + L + L + L = 4L$.

square

A rectangle is similar to a square. Unlike the square, however, two of its sides are longer than the remaining two sides.

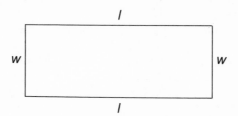

The measurement of the longer sides is called the *length* of the rectangle, and the measurement of the shorter sides is the *width*. Thus, its perimeter is

$$l + w + l + w = l + l + w + w = 2l + 2w.$$

Next we consider an everyday problem involving perimeters.

Example 1: A rectangular window is 12 inches wide and 20 inches high. A thin band of weather stripping is to be placed around the perimeter of the window excluding the bottom opening. The weather stripping costs 8¢ per inch. What will be the total cost of the weather stripping?

Step 1   Read and Represent

This problem is one in which a picture will be helpful. The window has the shape of a rectangle, and the weather stripping is to go around the window, except for the 12 inches at the bottom opening. I drew a simple diagram to denote this situation:

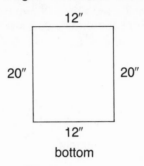

bottom

I have chosen the letter w to denote the length of weather stripping needed. This problem involves two relationships. First I must determine the distance around the window that is to be covered by the weather stripping. This distance will be the perimeter minus 12 inches for the bottom opening. Hence, w will be the sum of the other three sides, or $20 + 12 + 20$ inches, or 52 inches. The second relationship I need to represent is the cost of the weather stripping. If the weather stripping costs 8¢ for one inch, then the total cost will be $8w$, or $8 \cdot 52$, or 416¢ $=$ \$4.16.

Helpful hint: In this example I have eliminated Step 2 and have actually combined my translating and solving steps with Step 1. After drawing the diagram and establishing the relationships involved, this problem was fairly easy to solve. I had to do only two short calculations. This happens frequently with math problems. Be flexible! You do not have to include *every* step of your problem-solving method for *every* problem. If skipping a step or combining two or more steps allows you to keep your train of thought, by all means do so!

TRIANGLES AND THE PYTHAGOREAN THEOREM

Many practical problems can be related to the length of the sides of a particular geometric figure, the *triangle*. As I indicated earlier, the triangle is a polygon with three sides, labeled a, b, and c below. The triangle below is special in that sides a and b are perpendicular to each other (i.e., one line, a, can be considered vertical, and *b* must be horizontal with respect to *a*. When this occurs, the angle formed by the two sides is called a *right angle,* and the entire triangle is called a *right triangle.*

Long ago the Babylonians discovered that right triangles had a special property. This property was finally recorded by Greek mathematicians and is called the *Pythagorean theorem,* after Pythagoras, an early philosopher. It states that if you have a right triangle, in which *c* is the length of the longest side and *a* and *b* are the lengths of the remaining sides, the following equation is true:

$$a^2 + b^2 = c^2$$

(You will want to recall that x^2 denotes $x \cdot x$.) Any triangle satisfying the above equation is a right triangle.

The following everyday problem makes use of the Pythagorean theorem.

Example 2: What is the length of a ladder that is placed against a wall such that it is 9 feet off the ground and 12 feet away from the wall.

Step 1   Read and Represent

In doing this problem a diagram will again be very helpful. In the diagram you can see that the ladder, its height off the ground, and its distance from the wall form a right triangle with the corresponding dimensions. I chose *L* to denote the length of the ladder.

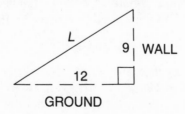

Step 2   Guess

Since L denotes the longest side, I could estimate that it is slightly longer than the run, maybe 14 feet. Again this step is helpful in that it prevents me from getting an answer such as 1,000 feet and not noticing that it is not a reasonable answer.

Step 3   Translate

By the Pythagorean theorem I know that the length of the longest side squared will equal $(9)^2$ plus $(12)^2$.
Thus, $9^2 + 12^2 = 81 + 144 = 225$.

Step 4   Solve

I must solve this equation:
$$L^2 = 225$$

There are many different ways to solve this equation. I chose to solve the equation by introducing square roots. The *square root* of a number, a, is another number, b, so that $b \cdot b$, *or* $b^2 = a$. For example, 2 is the square root of 4, since $2 \cdot 2$ or $2^2 = 4$. The square root of a is denoted as $\sqrt{a}$. Thus, $\sqrt{4}$, read as "the square root of 4," is 2. Although there do exist techniques for determining square roots (you may want to refer to an algebra text), most of the new hand-held calculators have a key to compute roots for you. Some square roots occur frequently and are easy to determine. For others, feel free to use a calculator. Remember that doing mathematics is using and applying math skills to solve practical problems. The actual calculations involved can often be done with a calculator. The hard part is getting to the calculations. If you have gotten that far in solving the problem, you deserve to have a little bit of the work done for you.
  Returning to the example, I must solve the equation $L^2 = 225$. I must find L, where $L^2 = 225$. That is, I must find L, where $L = \sqrt{225}$. The answer is 15, since $15 \cdot 15 = 15^2 = 225$.

Step 5   Check and Conclude

You can check that $15^2 = 15 \cdot 15 = 225$ is indeed equal to $9^2 = 9 \cdot 9$ plus $12^2 = 12 \cdot 12$; $225 = 81 + 144$.
  In my conclusion I must state that the length of the ladder is not just 15 but 15 *feet*.

The following practical problems also use the Pythagorean theorem:

1. If a 5-foot ladder is leaning against a wall and its top is 4 feet off the ground, how far away from the wall is the bottom of the ladder?
2. What is the length of wire needed to support a pole that is 15 feet high so that one end of the wire is attached to the ground 8 feet from the pole?

Answers are in Appendix B.

AREAS

The original problem is this chapter entails measuring the area of the floor in the dining area. What is area? *Area* is the measure of a region bounded by a closed curve. The process for determining this measure is similar to measuring the length of a line. You first need to find how many standard units fill up the figure.

In the rectangle below with sides that measure 6 units and 4 units, the area can be broken up into 24 squares, each measuring one meter on each side.

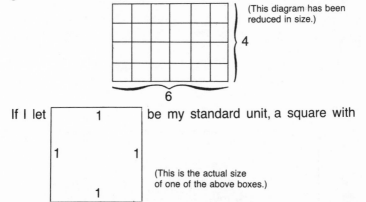

(This diagram has been reduced in size.)

If I let [1] be my standard unit, a square with

(This is the actual size of one of the above boxes.)

each side measuring one inch, I can see that the area of my rectangle will equal 24 of these standard units. This standard unit of measure is called a *square inch.* If you are working with meters, the standard unit will be a *square meter,* corresponding to a square in which each side has a length of one meter.

The area, *A,* of a rectangle can be determined by multiplying the length times the width. Thus, in our example the area is (6 inches) times (4 inches) = (24 inches)^2 = 24 square inches.

In general, if a rectangle has length, *L,* and width, *W,* its area will be

$$A = L \cdot W.$$

A large number of practical problems are concerned with measuring the area of a geometric figure. Let us return to the original problem in this chapter:

> Suppose that the dining area of your apartment is 6 feet wide and 10 feet long. You want to buy wall-to-wall carpeting for this area. How much will it cost if the carpeting is $12 per square foot?

Step 1   Read and Represent

After rereading this problem you should begin to realize that there are two unknowns—the area to be covered and the cost of carpeting that same area. The area, *A,* you now know is $L \cdot W$; *L* is the length of the dining area, and *W* is its width. The cost of the carpeting, which can be represented by *C,* is found by forming the product 12*A.*

I have again chosen to skip Step 2. Also, I have actually combined Steps 1 and 3. These steps can be summarized by the following:

(i)   *A,* area of dining room, equals $L \cdot W$.

(ii) *C.* cost of carpeting, equals 12*A.*

Step 4   Solve

Solving this problem will just be a matter of doing the two calculations for *A* and *C.*

In our example $L = 10$ and $W = 6$. The area becomes (10 feet)(6 feet) $= 60$ square feet. The cost of carpeting this area is $12*A,* or ($12)(60), or $720.

Step 5   Check and Conclude

I have chosen to leave the checking of these calculations to you. However, it is important to point out that the conclusion to this problem is *not* the area of the dining room. It is the *cost* of the carpeting that you are to find. It is often easy to get caught up in your solution and not give the answer asked for!

The practice problems below are all problems involving the concept of area:

(1.) Find the area of a square with each side measuring 9 centimeters.

(2.) Find the area of a rectangle that is 3 feet by 12 feet.

(3.) How many square feet of carpet will Ann need to carpet a rectangular floor that measures 11 feet by 14 feet?

(4.) If one pound of fertilizer is needed to cover one square foot of ground in a new garden, how many pounds are needed to cover a rectangular garden that is 12 feet by 6 feet?

(5.) What is the cost of paving a city playground measuring 50 yards by 35 yards if the paving costs $3.50 per square yard?

Helpful hint: This chapter is an excellent one to give you an opportunity to form some of your own problems and solve them. You might make up area or perimeter problems and exchange them with someone else who wants to review. This person could be a member of your family or a friend or co-worker.

The problems you each create are often really interesting and creative. In addition, you gain a lot from explaining problems to others, correcting their solutions and, of course, trying to solve their problems. Everyone in the Math Without Fear program found this technique to be fun and very helpful.

The more you enjoy doing math, the more math you will want to learn. Obviously, self-confidence will build, and anxieties will naturally fade away. If you can make doing math problems more pleasant by sharing ideas with someone else, you are sure to get good results.

REFERENCE

[1]Hackworth, R., and Howland, J. *Introduction to College Mathematics*. Philadelphia: W. B. Saunders, 1976.

CHAPTER 9
COPING WITH STATISTICS

> There are three kinds of lies: lies, damned
> lies, and statistics.—Benjamin Disraeli
> Get the facts first and then you can distort
> them as much as you please.—Mark Twain

Statistics? Who needs it? The word *statistics* probably is second only to the word *mathematics* in its ability to evoke negative reactions. Many adults view statistics unfavorably, and perhaps with good reason. Almost everyone, at one time or another, has been confused or misled by a collection of so-called valid statistics.

One day the latest statistics indicate that we should eat less beef. The next week, statistical reports indicate that iron deficiencies exist and we should eat more spinach! One set of statistics indicates that consuming any amount of saccharine is harmful, and the next set concludes that there is no danger.

The morning newspaper poll reveals that the economy is improving, but the evening edition reports that inflation is at an all-time high.

Two different manufacturers each claim to have the number-one pain reliever.

Almost every car dealer has the top-selling automobile.

Advertising agencies, politicians, government officials, stockbrokers, and salespeople are only a few of the groups of people who use statistics to convince us that theirs is the best product, the best platform, or the only solution to the energy crisis.

The list could go on and on of those who unintentionally and intentionally use statistics to confuse or mislead.

The quantity of numbers and statistics that we must deal with is increasing by leaps and bounds.

We are faced with statistics on TV and radio, in the newspaper, at the market, on the job, and at football games.

Who is leading in the polls?

What is our economic forecast for the future?

What is the average number of yards gained in the first quarter of the Redskins-Colts game?

Answering all these questions requires numbers and the basic tools of statistics. However, no one can deny the need for statistics. The ever-increasing collections of data—births, deaths, divorces, weather forecasts, automobile accidents, faulty items on a production line—are but a few of the many areas of everyday experience involving large quantities of information. The success of *The Book of Lists* is another indication. The science of statistics is used to analyze this information. The jobs done by economists, census takers, sociologists, life insurance companies, football teams, and, of course, mathematicians, all require statistics. After analyzing data, conclusions are drawn. The economy is unstable; 35 percent of city populations have moved to the suburbs; this year's team is making more hits on the average than last year's team.

The use of statistics goes beyond drawing conclusions. In everyday situations we often make important decisions based on conclusions derived from statistics: which car to buy, which candidate to vote for, which insurance policy to buy, what career change to make, what to eat, where to live, how to invest money.

All of these decisions are greatly influenced by statistics and by conclusions drawn from these statistics.

The fact that statistics can be misleading and misinterpreted is disconcerting.

As adults many of our important decisions are based on statistics. How can we defend ourselves?

Most adults do not really understand statistics; they are unfamiliar with its language and terms and do not have the tools or confidence needed to question the claims and conclusions of those who use statistics.

These characteristics are particularly indicative of math-anxious adults. The math-anxious have little faith in their abilities with fractions, let alone statistics.

As I said earlier, the word *statistics* is as frightening as the word *mathematics*. *Average, mean,* and *median* are terms many adults never have really understood. The reasonings behind the statistics and their conclusions are confusing. If people really do not understand statistics, how can they question the conclusions of statisticians or of some firms when the processes involved are not at all clear? As a result adults are forced to rely on the good judgment, honesty (?), and expertise of others who claim they can understand and interpret statistics.

This has two especially negative effects. First, it adds one more area in which you are dependent on the abilities of others. Self-reliance is decreased along with self-confidence. Second, you are placed in a vulnerable position. You have to put complete faith in the intentions and abilities of people who want you to buy their products, use their services or vote for their candidates, and so on. Unfortunately, not everyone has your best interest in mind. Desire for greater sales, bigger profits, and more votes often takes precedence over accuracy in reporting details, including all *pertinent* information.

Every adult should be wary of opinions, conclusions, and predictions made on the basis of a given set of statistics.

Math-anxious adults in particular need to become acquainted with the elementary concepts of statistics. They need to be alert to the possible misuses of statistics. They need to be able to read and evaluate newspaper and magazine articles. Finally, they need to be able to draw their own conclusions and make their own decisions based on what they have read. If you have been bewildered by the vast quantity of numbers filling all aspects of your life, you will want to pay close attention to this chapter. This chapter will explain some of the most common expressions used in statistics. I will talk about what the science of statistics really is, and I will describe some of its methods and processes. Finally, I will point out how statistics can be misinterpreted and misused and how to evaluate the conclusions of others.

After reading this chapter you will be able to deal with that vast quantity of numbers more confidently. You will be able to read and evaluate articles and claims with a critical eye. And, most important, you will feel more self-confident about your ability to function effectively as a consumer, voter, and as a self-reliant adult. Important decisions will be based more on your own conclusions and evaluations and less on the sometimes questionable claims of others.

WHAT IS (OR ARE) STATISTICS?

This question indicates that the word *statistics* can be interpreted in two different ways.

First, the word *statistics* can mean "actual figures or numerical facts."

For example, a listing of the average daily consumption of cigarettes by 200 teenagers would consist of 200 numbers (probably in terms of packs), or 200 statistics.

On the other hand, the word *statistics* can also mean "the study of figures or data," just as the word *mathematics* refers to a subject for study.

No wonder statistics can be confusing.

Even the word has two different meanings.

In the second sense, statistics consists of collecting, organizing, and interpreting data or information.

Statistics are also used to make conclusions and predictions based on the information gathered. For example, the *Literary Digest* predicted that Alf Landon would defeat Franklin Roosevelt in the presidential election of 1936, after a poll of ten million people. Only two million ballots were returned. The prediction, of course, turned out to be dead wrong. The *Digest*, which folded in 1937, had sent ballots to people listed in telephone directories and to registered automobile owners. In 1936 those who could afford a telephone or car did not represent the average voter. Instead, they represented a prosperous section of the voters, who tended to vote Republican.

This is one well-known example of how statistics can be misinterpreted or biased. I will refer to other examples of statistical misuse later. However, this example *does* reveal how the science of statistics is used to make predictions about a large class of objects.

CENTRAL TENDENCY, OR BEING IN THE MIDDLE

Statistics can be used to organize and present numerical facts or data.

For example, in a large corporation, a list of the salaries of its employees gives you information about the employees as a group. However, it might be interesting to determine the average salary of the employees, or perhaps the number of salaries being earned by women. This information gives you an idea of the trends of the figures, a measure of what statisticians call *central tendency*.

Central tendency is just an extension of the idea of averages. However, the word *average* in statistics can have more than one meaning. Let us take the following example.

In order to attract new employees a local construction firm advertises that the average salary of 50 employees at their firm is $10,000. The union claims that the average salary is only $8,500.

The 50 people employed at the firm include 48 whose net income is about $8,500. The remaining 2 hold managerial positions and make $46,000 annually. The first average, called the *arithmetic mean*, is found by adding the total income (i.e., 48 employees times $8,500 equals $408,000, plus the two receiving $46,000, or a total of $92,000 for a grand total of $500,000) and dividing by 50 to get $10,000.

The second average of $8,500 is the most common salary earned by workers at the firm.

This type of average is called the *mode* and represents the most common value or number occurring.

This case is just another example of how confusion and misunderstanding of math concepts are often the result of different meanings associated with certain terms and not the result of a lack of ability, or lack of a math mind.

The concept of average is one that is used constantly in everyday situations.

A business person wants to know the average return on his or her investments; car owners are concerned about the average mileage of their automobiles; a football team wants to calculate the average number of yards gained per game; the census bureau needs information on the average number of children per family. Averages are necessary in making many everyday decisions.

For example, if you have a car that presently gets, on the average, twelve miles to a gallon of gas, you may decide to trade it in for a smaller model with better mileage. A business person may decide to switch his or her investments from the stock market to mutual funds after comparing average returns.

The word *average* is one that you need to understand to be able to read and comprehend news articles, to make comparisons in shopping, to decide on career choices and even to choose leisure activities.

I have mentioned two kinds of averages so far: the *mean*, obtained by adding all of the values (test scores, weights, salaries, and so on) and dividing that sum by the number of test scores or weights that you added; the *mode* was the most common value.

One kind of average will be more useful than another, depending on the situation. In the case of the salaries of employees at the construction firm, the mean of $10,000 was just the average of the two people in management earning $46,000 with the much lower salaries of the remaining 48 workers. The mode actually gave more meaningful information by revealing that the most common or frequently paid salary was only $8,500. There is yet one more kind of average used in various surveys and studies. This average is called the *median*, which is the middle value of a set of numbers when they are put in increasing order.

In describing a set of salaries, the median gives you more information than the mean by indicating that half of the salaries are greater than a certain figure and half are less. For example, suppose a census of the incomes for 7 families reveals the following figures: 8, 7, 6, 8, 120, 7, and 5, where it is understood that these numbers stand for thousands of dollars. If we arrange this collection of data in increasing order we get 5, 6, 7, 7, 8, 8, 120. The median is $7,000. Half of the families make $7,000 a year or more, and the other half make $7,000 or less. But look at the mean!

The mean $= \dfrac{5 + 6 + 7 + 7 + 8 + 8 + 120}{7} = 23$. If we regard the mean as the average income, we would say that the average income is $23,000, when 6 out of the 7 families make $8,000 or less. Obviously, the high income of the one family has given us this high mean. This is a good example of how the different kinds of averages give different kinds of information.

One more note about medians. It is possible that you will want to find the median of an even number of values. For example: 2, 3, 4, 4, 7, 7, 8, 8. In this case there are two middle values, 4 and 7. To get the median, you take the arithmetic average of these two middle terms. The median is $\frac{4+7}{2} = 5.5$. Note that 5.5 is not one of the actual values, but it does represent the midpoint of the values when the values are put in increasing order.

In the practice problems that follow try to determine the mean, mode, and median of the sets of data:

(1) 1, 3, 11, 6, 8, 7, 6

(2) 11, 5, 3, 4, 3, 3, 2, 9

(3) 5, 5, 2, 3, 5

(4) 7, 7, 8, 8, 2, 2, 8

VARIATIONS OF DATA

Averages do not always give a true picture of the situation involved. Mathematics is, on the average, devoid of jokes.

There are, however, two very short stories that come close to being humorous.

The first story tells of the statistician who drowned in a lake with an average depth of three feet!

The second involves the mathematician with his head in the refrigerator and his feet in the oven who remarked, "On the average, I feel fine!"

In the case of the statistician, even though the lake had an average depth of three feet, in some places it was only a few inches deep, while in other places it was well over his head. And,

	mean	median	mode
(1)	6	6	6
(2)	5	3.5	3
(3)	4	5	5
(4)	6	7	8

Answers:

in the mathematician's case the hot temperature of the oven balanced with the cold temperature of the refrigerator has given an average temperature that would be quite normal.

Information about the amount of variation from the mean or average is sometimes needed to give an accurate picture. In some books this lack of information is called *disregarded dispersion* and is just another way statistics can be misused.

The next example also reveals how knowing only the mean can give you an incomplete picture.

Two different weight-reducing clinics each claim that their last class of 10 adults had an average weight loss of 12 pounds over a 6-week period. In choosing one group over the other you would assume that their success rates were about the same. However, let us take a closer look at their success with the figures below:

Weight-Loss Club A	*Weight-Loss Club B*
0	8
2	9
2	10
3	12
5	12
6	12
6	13
6	14
43	14
47	16
120	120

$$\text{mean} = {}^{120}\!/_{10} = 12 \qquad \text{mean} = {}^{120}\!/_{10} = 12$$

After inspecting the above figures I would choose Club B. Club B had a more consistent weight loss among all of the ten members. In Club A most members lost 6 pounds or less. Only the two dramatic weight losses brought the mean up to 12. Making decisions based only on the mean is sometimes unwise. More information, such as how far the data is spread out, is needed.

Reliable information on the scattering of data is extremely useful in business, government, and many practical situations. For example, a vitamin manufacturer wants to have a bottling process that will put, on the average, 50 tablets in each bottle. This could be done by putting 49 in one, 51 in the next, 48 in the third, 52 in the fourth, and 50 in the fifth . . . and would have an average of 50 tablets per bottle.

However, the bottling machine could also deposit an average of 50 tablets per bottle by putting none in the first, and 100 in the second, and so on! The manufacturer cannot afford this much variation in the bottling process. Customers receiving empty packages would naturally be upset! Any manufacturer is concerned with producing products of uniform quality, that is, they need to have little variability among the products. In the following sections I will discuss some of the basic methods used to measure this type of variability.

One way of indicating the spread of the data is to determine the *range*. The *range* is merely the largest value minus the smallest value. The following numbers are the number of children belonging to five different families: 2, 3, 4, 5, 6.

The range in this example would be $6 - 2 = 4$. The range often is used to report the change in stock prices over a period of time, and weather reports give the high and low temperature readings for the day.

The range is fairly easy to calculate, but it does not give a detailed picture of how the different pieces of data are scattered. In fact, it gives only information related to two values, the largest and the smallest. Suppose you had the following two sets of numbers: 2, 10, 10, 10, 10, 11, 11, 30   and   2, 2, 2, 2, 30, 30, 30, 30. Each set of numbers has a range of $30 - 2 = 28$. However, in the first case you might note that the majority of numbers are near 10.

In the second case, half of the numbers have the value 2 while the other half are 30. The second set has quite a wide dispersion of its numbers while the numbers in the first set are clustered fairly closely around the value 10. The range gives a very limited picture of the data involved. To remedy this situation mathematicians came up with other methods to indicate to what extent pieces of data are scattered, particularly in their relationship to the mean.

The second method of indicating the variation of data from the mean is called the *standard deviation*. Before discussing this new term, I want to introduce a new notation. The symbol $\bar{x}$ often denotes the mean. If x is one piece of data, then $x - \bar{x}$ will indicate how far away x is from the mean, $\bar{x}$.

This number, $x - \bar{x}$, is referred to as the *deviation* of x from the mean. Let us return to the example of the five numbers: 2, 3, 4, 5, 6.

The mean, $\bar{x}$, of these numbers equals

$$\frac{2 + 3 + 4 + 5 + 6}{5} = \frac{20}{5} = 4.$$

For each of the six figures you can calculate its deviation from the mean, 4. The deviations are listed below:

Data	Deviation from mean, $\bar{x}$
2	$2 - 4 = -2$
3	$3 - 4 = -1$
4	$4 - 4 = 0$
5	$5 - 4 = 1$
6	$6 - 4 = 2$

You might note that two of the deviations are positive and two are negative. When $x - \bar{x}$ is positive, that just corresponds to the fact that x is greater than the mean $\bar{x}$. For example, the deviation of 6 is equal to positive 2 since 6 is greater than the mean 4. In fact, 6 is 2 more than the mean 4. If the deviation is negative it means that x is less than $\bar{x}$. For example, 3 has a deviation of -1 since it is less than the mean 4.

When I defined the range, I actually was deriving a *single* number to describe how spread out the pieces of data are. In the second method, I also want to obtain a single number that describes the deviations. My first inclination would be to add all the deviations and divide by 5. I would obtain an *average deviation*. Look at what happens if I do this procedure. The sum of the deviations is equal to $(-2) + (-1) + 0 + 1 + 2 = 0$. The average of the deviations would be $^0/_5$ or 0. To get around this problem, statisticians work with the squares of the deviations. You will want to recall that the square of a number, x, is denoted by x^2, and means x times x.

The single number used to describe the variation, or spread, of the data is called the *standard deviation*.

It is the square root of the mean or average of the squares of the deviations. Let us return to our example. The next step needed is to calculate the squares of the deviations.

The information for our example is listed in the chart below.

Number of children	Deviations	Squared Deviations
x	$x - \bar{x}$	$(x - \bar{x})^2$
2	-2	4
3	-1	1
4	0	0
5	1	1
6	2	4
20		10

Mean, $\bar{x}$, $= \dfrac{20}{5} = 4$ Mean of the squared deviations $= \dfrac{10}{5}$

The standard deviation is the square root of the average (mean) of the squares of the deviations. It is the square root of $^{10}\!/_5 = 2$. The standard deviation is $\sqrt{2} = 1.41$. You might note that taking the squares of the deviations eliminated the problem of getting a total equal to zero. A summary of the steps needed to calculate the standard deviation is given below. Finding the standard deviation:

(1) Determine the mean, $\bar{x}$.
(2) List the deviations from the mean for each piece of data.
(3) Calculate the squared deviations, $(x - \bar{x})^2$.
(4) Obtain the mean of the squared deviations.
(5) Take the square root of item (4). This will be the standard deviation.

You will want to try your hand at calculating the mean and standard deviation of the sets of numbers below:

(1) 3, 2, 8, 10, 7
(2) 3, 4, 5, 2, 4, 7, 9, 6
(3) 2, 1, 0, 0, 5, 7, 6

	Mean	Standard Deviation
(1)	6	3.03
(2)	5	2.12
(3)	3	2.73

Answers:

The answers are approximate ones. You will probably find a calculator helpful.

Before returning to the misuse of statistics there is one more aspect of the standard deviation that you will want to consider. In the answers to the previous problems you probably noted that in some cases the standard deviation was larger than in others.

What does this larger standard deviation imply?

If the standard deviation is a larger number, this number implies that the data are more spread out. A small standard deviation indicates that the data are clustered more closely around the mean.

FREQUENCY DISTRIBUTIONS

Most of the examples so far have dealt with fairly small collections of data. However, everyday situations often involve large quantities of numbers. Organizing these larger sets of data could be a problem. One method of handling this problem is to use a *frequency distribution*. A frequency distribution will show how often each different piece of data occurs.

To illustrate the use of frequency distributions I have chosen an example involving a topic familiar to any student: test grades. Suppose the following list represents 20 grades on a test given to a class reviewing algebra: 80, 95, 75, 75, 65, 85, 85, 70, 90, 90, 95, 80, 80, 85, 70, 70, 100, 75, 85, 75.

To organize these data into a frequency distribution you first list each grade from smallest to largest in a column. Then you place the number of times that grade appears in the original list to the right of that grade. This number is referred to as the *frequency* of that particular value. This information corresponding to our example is given in the frequency distribution table below:

Grade	Frequency
65	1
70	3
75	4
80	3
85	4
90	2
95	2
100	1

The frequency table can also be used to take shortcuts in calculations such as determining the mean. You will recall that the mean is:

$$\frac{\text{add the individual pieces of data}}{\text{total number of pieces of data}}$$

Note that the sum of the individual pieces of data will be 65 + 70 + 70 + 70 + 75 + 75 + 75 + 75 + 80 + 80 + 80 + 85 + 85 + 85 + 85 + 90 + 90 + 95 + 95 + 100. (I have rearranged the order of the numbers; this will not change the sum.) This sum is also equal to 65 + 70(3) + 75(4) + 80(3) + 85(4) + 90(2) + 95(2) + 100 = 1,625. (You may want to recall that $x + x + x = 3 \cdot x$, or $x \cdot 3$.)

The second sum was obtained by multiplying each grade by its frequency and then adding all of these products. Also, if we add all of the frequencies, we are really counting the number of pieces of data: i.e., 1 + 3 + 4 + 3 + 4 + 2 + 2 + 1 = 20. The mean = $\frac{\text{sum of the } x \cdot f's}{\text{sum of the } f's}$ where x is a piece of data and f is its frequency. Therefore, the mean of our example is

$$\frac{1,625}{20} = 81.25$$

HISTOGRAMS AND LINE GRAPHS

As in many other areas of mathematics, a picture illustrating the information in a frequency distribution would certainly give one more feeling for how the data are distributed. This can be done by means of a *histogram*. A histogram for our test grades example would list all of the distinct grades, from smallest to largest, on the bottom line. Then a bar is drawn. Its center is the grade, and its height is given by the frequency of the grade. Frequencies are listed on the vertical scale. Here is a histogram for our example:

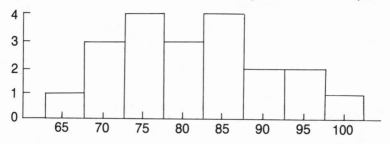

Another pictorial representation of data often used in statistics is the *line graph*. A line graph can be obtained from a histogram by placing centered dots on the tops of the vertical blocks. These dots, or points, are then joined by line segments. The line graph corresponding to our example is given below:

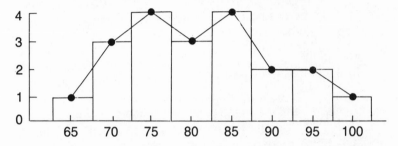

You may want to note that the line graph could have been obtained directly without drawing the histogram. You would merely place a dot over each grade so that its height corresponds to the frequency of the grade on the vertical scale. Then these dots are connected in the same manner.

NORMAL DISTRIBUTIONS

Many frequency distributions have line graphs that are approximately like the one below. The frequency distribution corresponding to line graphs such as this are referred to as *normal distributions*. Notice that the highest point of the line graph occurs in the middle. Also, the curve is bell-shaped and symmetrical around a vertical line drawn to the highest point. Therefore, if the left side of the curve were flipped over the vertical line, it would lie directly on top of the right side.

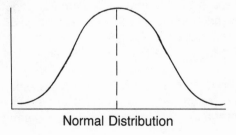

Normal Distribution

It turns out that the line graphs corresponding to many real-life distributions are approximately normal. Happily, for those using statistics, normal distributions have many nice properties. For one, in a normal distribution the mean, median, and mode all coincide. This implies that the average piece of data is also the one that occurs most frequently and is also in the middle when the numbers are listed in increasing order. In addition, mathematicians have been able to prove some useful results related to normal distributions and standard deviations. One of these results says that if a distribution is normal, then 68 percent of the numbers involved can be found within one standard deviation of the mean. Thus, 68 percent of the data will be between the mean minus one standard deviation and the mean plus one standard deviation. (Again I am using $\bar{x}$ to denote the mean and sd to denote a standard deviation.) This property is illustrated below:

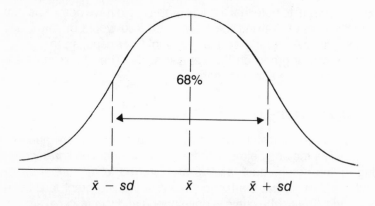

$$\bar{x} - sd \qquad \bar{x} \qquad \bar{x} + sd$$

In the next example you can combine this information with your past work on percentages to answer the question involved.

Suppose that you are told that the recorded gasoline mileage of 50 new cars in the fall formed a normal distribution. You are also told that the average gasoline mileage of the 50 cars was 19 miles per gallon, and the standard deviation was 6 miles per gallon. How many cars had mileage between 13 and 25 miles per gallon?

Since the different mileages obtained are normally distributed, you can picture this data in the diagram below, where the mean or average of 19 miles per gallon is in the middle.

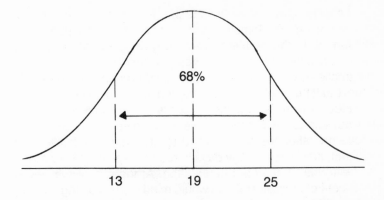

Note that the standard deviation is 6. From the previous section we now know that 68 percent of the numbers lie between 19 minus one standard deviation and 19 plus one standard deviation. This means that 68 percent of the data lie between $19 - 6 = 13$ and $19 + 6 = 25$. To answer our original question we calculate 68 percent of 50. Recalling techniques learned in Chapter 6, we find that 68 percent of 50 is determined this way:

$$\frac{68}{100} \cdot 50 =$$
$$\frac{68 \cdot 50}{100} =$$
$$\frac{3,400}{100} = 34$$

Summarizing, we would say that 34 cars will have gasoline mileage between 13 and 25 miles per gallon. Normal distributions are important in statistics since many everyday situations and their corresponding distributions seem to fall into this category. I have mentioned only a few basic ideas of normal distributions, but if you do any more work in statistics, it will be a topic that will receive much attention and, of course, will be dealt with in much more detail.

Helpful Hint: Many math problems, especially problems in statistics, can be intricate and sometimes very involved. Do not be afraid to work on a problem for a short period, leave it to do something else, and then return to it later. One of the nice aspects of math problems is that they will not change while you are gone! Sometimes just doing something else and returning to the problem later can give you a fresh outlook. I recall a stumbling block I had with my Ph.D. dissertation. I spent an entire week on a particular problem and seemed to be getting nowhere fast. I seemed to be going in circles. The more I worked, the more confused I became. Finally, my adviser suggested that I leave the problem and write up some of my other results. I wanted to stay with the problem, but took her advice. A few days later I met again with my adviser with the other work I had written. At the end of the session I asked my adviser if she would mind my explaining the problem one more time. To my own amazement, as I was explaining the problem I began to see a possible solution and solved it later that night. I would never have believed that leaving the problem to do something else could give me fresh insights when I returned to it! You may also find the technique helpful. There is certainly no rule that states that you must solve a math problem in one sitting!

In this chapter I have introduced just a few of the basic terms and concepts used in statistics. A full treatment of the elements of statistics would require an entire book. However, this chapter will make you feel more comfortable with the language of statistics and some of its methods.

In addition, I have also included some of the ways statistics can be misinterpreted or can be used to mislead you.

I will devote the rest of this chapter to further examples of possible misuse of statistics. These examples will help you become more aware of loopholes in claims made by manufacturers, politicians, and a host of other individuals you encounter daily.

FURTHER MISUSES OF STATISTICS

Statistical Graphs

Statistical graphs can be misused and can give a misleading picture of the data involved. Being aware of these possible misuses will help you to evaluate magazine reports, books, and newspaper articles with more confidence. The two line graphs below show the profits obtained by a local grocery merchant for each month of the past year. Months, from January to December, are listed on the bottom line. The profits, in terms of thousands of dollars, are listed along the vertical line.[1]

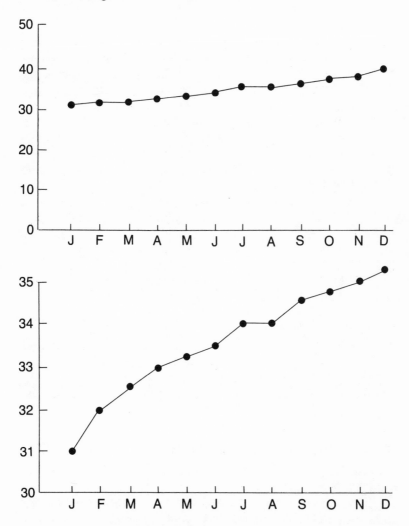

Both graphs provide the same information, but the second graph certainly seems to show that profits are increasing at a great rate, and it is more impressive. The scale has been changed so that it starts at 30 and the numbers on the vertical scale are spread much farther apart. Be careful of how scales are drawn when you interpret statistical graphs.

Insufficient Information

I already have referred to examples of misused statistics involving different kinds of averages and also misuses that occur by disregarding the variations among the numbers even when you know the average (you may recall the poor statistician who drowned in the lake with an average depth of three feet). Statistics can also be misleading when pertinent information is not included.

A classic example of such cases is a survey that was made some years ago at Johns Hopkins University. In that survey it was noted that one-third of the women attending the university had married faculty members.

Naturally, this information seemed to imply that if a woman went to Johns Hopkins, the prospect of marrying one of her professors seemed very good! However, the survey failed to mention that there were only three women enrolled at the time and, indeed, one had married a faculty member.

In another case, it was reported that the latest statistics revealed that cases of influenza and pneumonia were practically confined to three southern states. (Naturally, this information would not help the tourism industry for the three states!) What the report failed to include was the fact that the reason the reported cases of influenza and pneumonia were restricted to these states was that these states had laws requiring all such ailments to be reported. All other states had long stopped requiring such reporting.

EVALUATING AND JUDGING STATISTICS

In developing methods to judge the validity of statistics you will need to ask yourself a series of questions. The following questions are similar to the ones suggested by Darell Huff in his popular book "How to Lie with Statistics" (see bibliography).

(1) Who supplied these statistics?
Sometimes different groups will provide only information that

supports their particular interests. This may account for discrepancies between the oil companies' statistics and the government's, for example. Be aware that the study may be biased.

(2) Where is the evidence?
(3) What information has been neglected?
Keep an eye out for missing information! Sometimes essential items such as the kinds of averages used, the size of the samples gathered, and other methods or facts may not be reported. Often it is this missing information that is needed most to evaluate the statistics.

REFERENCE

[1]Huff, Darrel. *How to Lie With Statistics*. New York: W. W. Norton, 1954.

CHAPTER 10
TAKE A CHANCE– ODDS AND PROBABILITIES

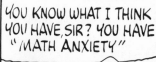

"I probably will not be able to go to the next meeting."

"It is likely that taxes will go up next year."

"My chances for a promotion seem slim."

"The odds for winning the lottery are definitely not in my favor."

"There is a one-in-three chance that the next adult you meet has math anxiety."

All of the statments on page 182 have one thing in common. They all share an element of uncertainty. Not one of the sentences states that something is certain. They only imply that it is likely that taxes will increase, or that a promotion is not likely, or that your next acquaintance probably will be math-anxious.

If you wrote down everything that you said in a single day, you would probably have included various statements similar to the ones above. (You might note that even that sentence has some uncertainty to it!)

Everyday situations are filled with uncertainties. Many times it seems as if there are fewer things we can be certain of than those that are uncertain.

I have a job today, but can I be certain I will have it next year?

Will my salary keep up with inflation?

Will there be enough heating oil for the winter?

Will it really snow tomorrow or is the weather report wrong again?

The list could go on and on, filled with items of uncertainty. Naturally, uncertainties such as these can lead to fear and anxiety. How can you make decisions if you cannot be certain of the future?

Probability is the mathematician's answer to such a question.

Probability has to do with things that *might* occur.

It tries to represent the likelihood that a certain event will occur by assigning a number to represent this likelihood.

For example, if you flip a coin, it is equally likely that you will get heads or tails. In other words you have 1 chance out of 2 for the coin to land heads. Therefore, you can assign the number 1/2 to denote the probability of getting heads.

When you think of the word *probability*, or *odds*, one of the first things that comes to mind is gambling: Atlantic City gambling casinos, betting on the horses at Belmont, bets made on boxing matches, football games, and the World Series, and tickets bought for state-operated lotteries. Probability had its origins with games of chance.

The uses of probability today are not restricted to gambling and games. Probability is used daily to make decisions in business, government, and industry.

As adults we also must be able to think in terms of probabilities. Various decisions we make must be based on the likelihood of certain events occurring.

These decisions include investing money, buying stock, and making major purchases. Women, in particular, as more and more are entering the job market, find that they must make decisions based on the ideas of probability. In my own case it was necessary to make a decision regarding retirement. In signing up for the retirement fund I had to decide whether I wanted 50 percent of the money invested in stocks and 50 percent in bonds or maybe 75 percent in stocks and the remaining 25 percent in bonds. I had to weigh the possibilities of the stock going down in value against the possible higher return if it should rise. Maybe it would be safer to put most of the money in bonds. I am sure you also could think of several situations in which you have had to make a decision by weighing all possibilities and then choosing the one you felt was most likely. Many of us do not feel comfortable when we have to make such decisions—when we have to take a chance. Often figures and percentages and probabilities are involved, and we really do not have a feeling for what they mean.

This is particularly true of the math-anxious adult, who does not like numbers and calculations anyway.

My aim in this chapter is to make you feel more comfortable with some of the basic terms and ideas connected with probability—what it is and what it is not.

PROBABILITY—A MEASURE OF CHANCE

Probability is a measure of chance. It measures the likelihood that a certain event will occur. Suppose you had an ordinary die on which each of the faces has from 1 to 6 dots. When you roll the die I think you would agree that it is equally likely that the die will land with any one of the 6 sides facing upward. What is the chance of rolling a 5? There is 1 chance out of 6 for the die to land with the 5 on top. You might say that a 5 has a 1-in-6 chance of being tossed. We say that the probability of tossing a 5 is 1/6. The mathematical symbol for this probability is given below:

$$P\,(5)\ =\ 1/16$$

Let's take another example. Suppose you have a well-shuffled deck of 52 playing cards, and you are asked to select one card. What is the likelihood of your selecting the queen of hearts? Since there is only one queen of hearts, and you would agree that each of the 52 cards has an equal chance of being chosen, you could say that it has a 1-in-52 chance. You would say that its probability is 1/52. Let's go a step further. Suppose you wanted to know the probability that your chosen card was a heart. What would its probability be? There are 13 hearts in a deck of 52 cards. Therefore, you would have 13 chances out of 52 for drawing a heart. Conclusion? The probability of getting a heart is 13/52, or P(heart) $=$ 13/52

Our two examples suggest the following definition of probability:

$$\text{Probability of an event} = \frac{\text{number of favorable outcomes}}{\text{total number of outcomes}}$$

At this point it will be helpful to continue building our math vocabulary by defining what we mean by such words as *event*. In tossing the die you want to observe whether a 1, 2, 3, 4, 5, or 6 has been tossed. This process is called the *experiment*. The possible results, in our case 1, 2, 3, 4, 5, or 6 are referred to as the *outcomes*. An *event* is a particular subcollection of the possible outcomes. Drawing a heart from the deck of 52 cards was an event since it represented 13 of the possible 52 outcomes when selecting a playing card from a well-shuffled deck.

Returning once again to our example of tossing the die, what would be the probability of tossing a 7? Naturally, your response would be that this is an impossible event. There is no way to toss a 7! The probability of tossing a 7 $= P(7) = \%_6 = 0$. This holds true in general: the probability of an impossible event is always zero!

Now suppose I asked you what is the probability that you will roll a number between 1 and 6 when you toss the die. This time your response would be that getting a number between 1 and 6 is a certain event that can occur in 6 different ways. The probability of getting a 1, 2, 3, 4, 5, or 6 $= P$ (getting a number between 1 and 6) $= \%_6 = 1$. In general, the probability of an event that is certain is 1. Since the number of favorable outcomes for an event is always less than the total number of outcomes the probability of an event will always be a fraction between 0 and 1 Note that $0 = \%$ and $\%6 = 1$ both can be considered as fractions.

Throughout this chapter I am going to assume that in a given experiment all of the outcomes are equally likely, that is, each outcome has an equal chance of occurring. There will be no loaded dice, no stacked cards, and no weighted coins. What does this imply for our definition of probability? Suppose an experiment has n equally likely outcomes. Furthermore, suppose that an event, E, corresponds to m of these outcomes. Then

$$P(E) = \frac{\text{number of favorable outcomes}}{\text{total number of outcomes}} = \frac{m}{n}$$

In tossing a die, suppose you were interested only in whether you tossed an odd number. Three outcomes correspond to this event, namely 1, 3, and 5. Thus,

$$P(\text{odd number}) = \frac{3}{6}$$

Suppose you are playing a game and your opponent is interested in the case where you do not get an odd number. If E is the event corresponding to getting an odd number, your opponent would likely want to know $P(\text{not } E)$. There is a notation to represent "not E"—$\bar{E}$. How many outcomes correspond to "not an odd number"? There are three: 2, 4, and 6. Thus, $P(\bar{E}) = \frac{3}{6}$. Note that this is $1 - P(E)$. This is always true. Going back to our original example, the probability of not getting the queen of hearts is equal to $\frac{51}{52}$, which is $1 - \frac{1}{52}$.

Sample space is another term used in probability. A sample space for an experiment is simply the set of all possible outcomes for the given experiment and is denoted by S. Thus, the sample space, S, for tossing the die is $S = 1, 2, 3, 4, 5, 6$. Sometimes determining the sample space for an experiment is not that easy, especially if the experiment involves two or more steps. Fortunately, there is a method that can help you make sure that no outcome has been left out in your final listing. The method is drawing *trees*, (not the maple and oak variety!), or *tree diagrams*. A tree diagram begins at a point that represents the beginning of the experiment. Next, you draw a number of branches corresponding to the number of outcomes in the first step. At the end of each of these branches you then draw branches corresponding to the different outcomes at the second step. This process continues until all steps have been completed.

In the tree diagram below I have been able to represent the sample space for the experiment of tossing a coin twice. By beginning at the initial point of the tree and following each of the distinct paths in the tree you can obtain a

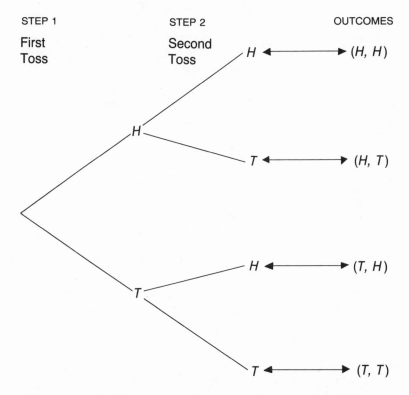

STEP 1   STEP 2   OUTCOMES

First Toss   Second Toss

H (H, H)

H

T (H, T)

H (T, H)

T

T (T, T)

listing of all possible outcomes. These outcomes are listed on the right. The sample space for tossing a coin twice consists of four pairs: the first entry in each pair denotes the result of the first toss and the second entry denotes the result of the second toss of the coin. Therefore, $S = \{(H, H), (H, T), (T, H), (T, T)\}$.

In the above example the coin is assumed to be a balanced coin, and each of the four outcomes is therefore equally likely. What is the probability of getting 2 heads? If E denotes the event, two heads, then there is only one outcome in S corresponding to E, the outcome (H, H). Since there are four equally likely outcomes corresponding to this experiment, the probability of E, $P(E)$, is equal to $\frac{1}{4}$.

Now suppose you had wanted to determine the probability of getting exactly 1 head. What would be the probability of this event? You will have to be a little careful with this case. What are the outcomes corresponding to the given event? Only two outcomes, (H, T) and (T, H), correspond to the event of getting exactly 1 head. Note that (H, H) does not correspond. Again, recalling that each of the 4 possible outcomes is equally likely, we can say that P (exactly 1 head) = $\frac{2}{4}$.

The next example is similar to the one above. The number of outcomes in the sample space of this experiment is larger. A tree diagram is particularly useful in obtaining a complete listing for S. This experiment consists of tossing, not a coin, but a single die twice. (This would be the same as tossing a pair of dice once.) How many different outcomes would there be? The tree diagram opposite reveals that there are 36 different possible outcomes corresponding to tossing a die twice.

The example on page 189 seems to suggest that there is a method of finding the total number of outcomes without first obtaining a complete listing of S. You might note that at step 1 there were 6 different possible outcomes. Corresponding to each of these 6 outcomes there were 6 more possible outcomes at step 2. There was a total of $6 \cdot 6 = 36$ possible outcomes. This idea is extended in what is called the *multiplication principle.*

MULTIPLICATION PRINCIPLE— DETERMINE THE TOTAL NUMBER OF OUTCOMES

If an experiment consists of n different steps to be completed in succession and there are x_1 ways to do step 1 (the subscript $_1$ is used to indicate that this number corresponds to the first step), x_2 ways to do step 2 ... and x ways to do step n, then the total number of ways to do the entire experiment is the product $[x_1 \cdot x_2 \ldots xm]$.

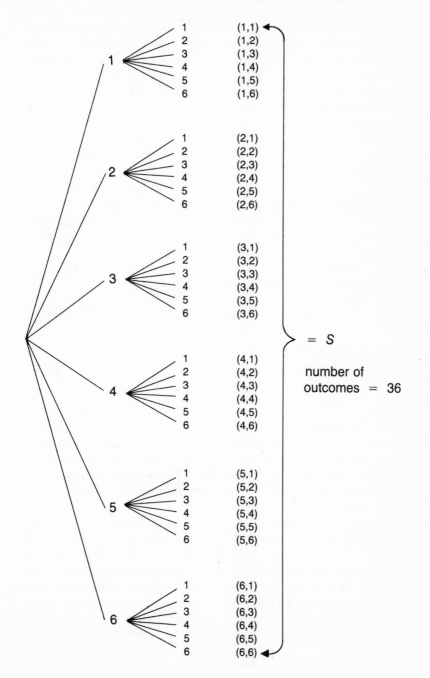

Let us use the multiplication principle in determining the number of outcomes for an experiment that consists of flipping a coin and then tossing a die. There would be 2 outcomes corresponding to the first step and 6 different outcomes corresponding to the second step. The net result is a total of $2 \cdot 6 = 12$ outcomes. For a complete listing of these outcomes you can refer to the tree diagram below.

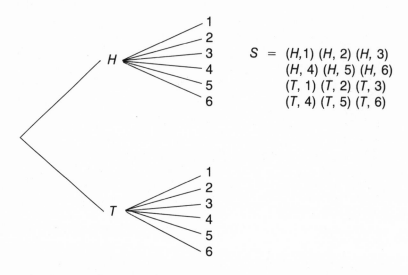

$$S = (H,1) \ (H, 2) \ (H, 3)$$
$$(H, 4) \ (H, 5) \ (H, 6)$$
$$(T, 1) \ (T, 2) \ (T, 3)$$
$$(T, 4) \ (T, 5) \ (T, 6)$$

The multiplication principle can also be used in determining the number of ways of arranging a given set of objects. Suppose you actually had not 9 but 5 different books and you wanted to calculate the number of ways these 5 books could be arranged on a shelf. How many different arrangements are possible? We might think of one arrangement of the five books as an experiment, or process, that involves 5 steps. Each step consists of assigning one of the books to one of five spots. Since there are 5 books to begin with, there are 5 ways to make the first assignment, or 5 ways to do step 1. Having chosen a book for the first spot there remain 4 books available for the second spot or 4 ways to do step 2. Now there are 3 books left to choose from for the third spot, 2 books left for the fourth spot, and finally 1 book left for the last spot. Thus, there are 5 ways to do step 1, 4 ways to do step 2,

3 ways to do step 3, 2 ways to do step 4, and 1 way to do step 5. Using the multiplication principal this implies that there are $5 \cdot 4 \cdot 3 \cdot 2 \cdot 1 = 120$ ways to do all 5 steps. That is, there are 120 ways to arrange 5 books on a shelf. You might note that there is a pattern of numbers involved here. The number of arrangments of 5 books is represented by taking the product of 5 and all the whole numbers less than 5 in decreasing order. A special mathematical symbol is used to denote such a product. It is called a *factorial*, and the product is denoted 5!. For example, $8! = 8 \cdot 7 \cdot 6 \cdot 5 \cdot 4 \cdot 3 \cdot 2 \cdot 1 = 40,320$.

In general, the number of ways of arranging N objects is equal to $N!$ and is called a *permutation* of the N objects. Permutations form a major topic in probability, and if you continue with your development of probability you will see a lot more of permutations.

INDEPENDENT AND DEPENDENT EVENTS

It is possible to extend the basic rules of probability to more complicated problems. In everyday situations one is often concerned with series of events that occur in succession. For example, in playing a game of chance you might be interested in the outcome of tossing a coin 10 times in a row, or maybe in the result of drawing 2 cards from a shuffled deck of ordinary playing cards.

Suppose you draw one card from a well-shuffled deck. What is the probability that this card will be a club? I know that there are 13 clubs in the deck: the 2, 3, 4, 5, 6, 7, 8, 9, 10, jack, queen, king, and ace of clubs. There are 52 cards in all. From the previous sections we can therefore determine that

$$P(\text{card is a club}) = \frac{13}{52} = \frac{1}{4}.$$

In other words, there is one chance in four of drawing a club. Now suppose that the selected card is put back into the deck, which is then reshuffled. Suppose that a card is again selected. What is the probability that it will also be a club? Again, there are 13 clubs in a deck of 52 cards. This implies that the probability that the second card will be a club (given that the first card was replaced in the deck) is ¼. Now suppose that I had asked you in the beginning to find the probability that both cards drawn will be clubs. You can then extend the idea of the multiplication principle. How many favorable ways can the event, both cards are clubs, occur? There are 13 ways for the first draw and 13 ways for the second draw. By the multiplication principle there are 13 · 13 = 169 favorable ways for the event to occur. The total number of ways two cards can be drawn is 52 · 52 = 2,704. Hence,

$$P \text{ (both cards are clubs)} = \frac{\text{number of favorable outcomes}}{\text{total number of outcomes}}$$

$$= \frac{169}{2,704} = \frac{13 \cdot 13}{52 \cdot 52} = \frac{13}{52} \cdot \frac{13}{52}$$

Note that $^{13}\!/_{52} = P$ (first card was a club) and the other $^{13}\!/_{52} = P$ (second card is a club). We could rewrite the above information as P (both cards are clubs) = P (first card is a club) · P (second card is a club). The probability that both cards are clubs was found by multiplying the probabilities that the first card is a club by the probability that the second card is a club. Note that this second probability is based on the fact that the first card was replaced in the deck.

In general, the probability of several events happening in succession is equal to the product of the probabilities of each event happening.[1] Let us look at another example using this rule. Assume that an experiment consists of flipping a coin and then tossing a die. The tree diagram below shows the 12 different possible outcomes:

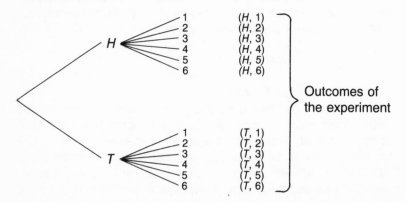

Now suppose that you wanted to determine the probability of getting heads on the coin and a 5 on the die. One method of finding this probability would be to list the tree diagram above; note that there are 12 equally likely outcomes of which $(H, 5)$ is one. Hence, $P(H, 5) = \frac{1}{12}$. As an alternate method I could have determined the probability of heads on the coin and 5 on the die without listing the tree diagram. From the previous section I should be able to determine the probability of $(H, 5)$ by multiplying the P(heads on the coin) by P(tossing a 5 on the die). The P (heads) $= \frac{1}{2}$ and P (5) $= \frac{1}{6}$. This implies that P $(H, 5) = \frac{1}{2} \cdot \frac{1}{6} = \frac{1}{12}$.

Let us now return to the playing card example. In this example the first card drawn was placed back in the deck before the second card was drawn. What occurred on the first draw had no effect on the result of the second draw. Events that have such a relationship are given a special name. Events that have no influence on each other are called *independent events*. Now assume that you change this example slightly. Two cards are still to be drawn from the deck. However, this time the first card selected is *not* replaced in the deck before a second card is drawn. Instead the first card is set aside. Now what is the probability that both cards will be clubs? In this case the result of the first draw will influence the probability that the second card drawn is a club. The probability that the first card will be a club is the same as before, $\frac{13}{52}$. However, this card is set aside. This means that there are only 51 cards remaining in the deck. If you assume that the first card was a club, then there are only 12 clubs remaining in the 51-card deck. The probability that the second card drawn will be a club is therefore $\frac{12}{51}$. Using the previous section on successive events, P(2 clubs being drawn in succession)

$$= \frac{13}{52} \cdot \frac{12}{51} = \frac{1}{4} \cdot \frac{12}{51} = \frac{1}{4} \cdot \frac{4}{17} = \frac{1}{17}.$$

In general, when events are influenced by other events they are called *dependent events*.

The next example will further illustrate the difference between independent and dependent events. On a table you have a box containing different colored balls: 4 black, 3 red, and 6 yellow. Two balls are to be drawn in succession, with the first ball being replaced before the second is drawn. What is the probability that the first ball will be black and the second ball will be red?

These events (first ball will be black and second ball will be red) are independent events since the first ball is replaced and does not influence the outcome for the second ball drawn. The probability of first a black ball being drawn is $= \frac{4}{13}$, since there are 13 balls and 4 of them are black. When the second ball is drawn, there are again 13 balls to choose from, of which 3 are red. This implies that P(second ball is red) $= \frac{3}{13}$. Hence, the probability that the first ball will be black and the second ball will be red is $\frac{4}{13} \cdot \frac{3}{13} = \frac{12}{169}$.

Now suppose the above example is changed to read this way: the second ball is drawn *without* the first ball being replaced. Now the event (first ball will be black) does influence the possible outcome that the second ball drawn will be red. Now there are only 12 balls to choose from, of which 3 are red. Therefore, P(second ball will be red) $= \frac{3}{12} = \frac{1}{4}$. The probability of getting a black ball on the first draw followed by a red ball on the second is $\frac{4}{13} \cdot \frac{1}{4} = \frac{1}{13}$. Notice that in this case the events (black ball on first draw and red ball on second) are dependent events.

PERCENTAGES AND ODDS

We already have seen that probabilities are merely fractions or ratios representing the likelihood of a certain event occurring. In this respect this chapter seems to relate to Chapter 5. Probabilities can also be expressed in terms of percentages and odds. Both Chapter 5, on ratios, and Chapter 6, on percentages, have an application to probability.

Suppose the probability of an event, E, is $\frac{1}{2}$. Since $\frac{1}{2} = \frac{50}{100}$, we could also say that the probability of E, $P(E) = \frac{50}{100}$. But $\frac{50}{100}$ is a percentage. The probability of E occurring could be restated by saying that E has a 50% chance of occurring, or $P(E) = 50\%$. In the particular case where $P(E) = 50\%$, the probability of E *not* occurring is $1 - \frac{50}{100}$, which is also equal to $\frac{50}{100}$, or 50%. The chance involved is $50 - 50$, a 50% chance that E occurs and a 50% chance that it does not occur.

From experience in reading sports columns and listening to newscasts you may also be familiar with another method for expressing probabilities—*odds*. If a sports column indicates that the odds that team A will win in the play-offs is 4 to 3, written 4 : 3, what probabilities are implied? The odds 4 to 3 mean that team A has 4 opportunities to win and 3 to lose for a total of 7 outcomes in the sample space. The probability that team A will win is $\frac{4}{7}$ when the odds are 4 : 3.

You can also start with the probability of an event, E, and then determine the odds in favor of E occurring. Suppose that the weather report predicts that the probability of rain tomorrow is $\frac{2}{10}$. What is the probability that it will *not* rain tomorrow? From earlier statements in this chapter you know that the probability of *not E* is $1 - P(E)$. The probability that it will not rain tomorrow is $1 - \frac{2}{10} = \frac{8}{10}$. The odds in favor of rain are 2 to 8, or 2 : 8. In general,

$$odds\ in\ favor\ of\ E\ = \frac{\text{number of favorable outcomes}}{\text{number of unfavorable outcomes.}}$$

Let us return once again to the tossing-a-die example. What would be the odds in favor of getting a 4 on 1 toss of the die? There is 1 favorable way the die can be tossed, and there are 5 unfavorable ways. This indicates that the odds in favor of tossing a 4 are 1 to 5, or 1 : 5.

As one last example connected with odds we might as well consider our now familiar coin-flipping experiment. If you flip the coin once, what are the odds in favor of the coin landing with heads up? There is 1 favorable way the coin can land and 1 unfavorable way. The odds, in favor of heads would be 1 to 1, or 1 : 1. You could then use the common expression, "the odds are even," to describe this situation.

MATHEMATICAL EXPECTATION

Probabilities are used in making decisions related to the payoff, or *mathematical expectation*, of an experiment. For example, if a local merchant is considering buying an ad on the radio, he or she must determine whether the amount of business the ad would bring in will exceed the cost of the ad. The amount of business brought in by the ad will be the payoff, or mathematical expectation, that is supposed to be greater than the cost.

We define the *mathematical expectation* of a given outcome as the product of its probability and the amount of the payoff.

Decisions on buying radio time are often based on statistically determined mathematical expectations. This example involves another type of probability called statistical probabilities. Statistical probabilities are determined by past experiences. For example, there is no way of knowing in advance how many people will purchase your product because of having heard your ad. However, from previous experience the probability that the ad will bring in extra business is $\frac{4}{7}$. Now suppose that the extra income from a successful ad (one bringing in extra business) is judged to be around $2,625. The cost of the radio ad is $1,200. Is the ad a good venture? The mathematical expectation would be $(\frac{4}{7}) \cdot \$2,625$, or $1,500. Thus, the ad is probably a good idea since it is expected to bring in $1,500 and costs $1,200 for a profit of $300.[2]

Perhaps this chapter has helped you become more familiar with some of the basic terms and tools of probability.

In this chapter and throughout the entire book my goal has been to help you become more at ease with mathematical language and symbols, to give you some of the reasons and logic behind various methods, to show you how to apply basic skills to everyday situations, and most of all to help you realize that past failures with math can be overcome by decreasing math anxieties and increasing math confidence.

I do not want to imply that math problems will never again be difficult for you or will never again cause some anxiety.

You probably will meet a problem that at first glance you cannot solve.

Mathematicians have this kind of experience all the time. And it is quite natural for some anxieties to begin to creep in. Having built up a degree of math confidence you will be able to keep these anxieties from overwhelming you and will be able to begin a plan of action for approaching the problem.

Like the many adults who participated in the Math Without Fear program at Georgetown University, you will find that in addition to gaining confidence in dealing with math you will also have gained pride and confidence in succeeding with something you could never do before.

Many in the program found that this new confidence carried over into other areas as well. You may consider returning to school, switching jobs, or even attempting to do your own car repairs or minor plumbing, or maybe you'll think about taking up a hobbie, such as photography. Whatever your interests or goals, mathematics need not hold you back.

Now that your math anxieties have changed to math confidence, the odds are really in your favor, no matter what your goal!

REFERENCES

[1]Jacobs, Harold R. *Mathematics, A Human Endeavor.* San Francisco: W. H. Freeman, 1970.
[2]Hackworth, R. and Howland, J. *Introduction to College Mathematics.* Philadelphia: W. B. Saunders, 1976.

APPENDIX A

ANOTHER LOOK AT ARITHMETIC

Although you use arithmetic every day in situations ranging from counting your change to estimating your income tax, there are certain arithmetic skills that can be troublesome. Many adults in the Math Without Fear program found that the arithmetic used now in algebra is slightly different from the arithmetic they are used to. The changes in arithmetic came about with the growth of the "new math." The actual processes are the same, but the names and methods are different. This is the major reason why parents find they cannot help their children with their homework! In this appendix I have included examples of topics that the adults in MWF found most helpful. It is not a complete review of arithmetic, but will reacquaint you with some basic skills and give you some of the reasons behind specific methods.

SIGNED NUMBERS

The term *negative number* may be new to you. However, the idea behind negative numbers is one you have probably used in various everyday situations.

If you gain 5 pounds and then lose 7 pounds, what is your net result? It would be a net loss of 2 pounds. To represent this net loss you can use this mathematical symbol: $^-2$. If you bet 10 dollars on a horse race and lose, you can represent this loss as $^-10$ dollars. If you want to indicate that the weather is cold and the temperature is 12 degrees below zero, the symbol $^-12°$ will give the desired representation. Mathematicians use this sign (⁻), to represent situations such as these. The symbol +4 is used to denote a gain of 4 pounds, net winnings of $4, or 4° above zero. A number with the plus sign (+) in front is called a positive number. If no sign is designated, for example, 7, it is understood to stand for the positive number, +7.

Mathematicians usually picture signed numbers as points on a line. (You may want to think of a thermometer turned sideways.) This picture is given below and is called the *real number line*.

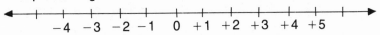

Each number corresponds to a point on the line and vice versa. The negative numbers are to the left of zero and positive numbers are to the right of zero. Notice that zero is neither positive nor negative. Since ⁻5 is the same distance from zero as 5, but is in the opposite direction, it is called the opposite of 5. Also, 5 is the opposite of ⁻5.

In the elementary grades you learned that numbers could be combined to get a third number by addition, subtraction, multiplication, and division. You can do the same with signed numbers, but there are a few rules you will need to follow.

If you lose 5 pounds and then lose 3 more pounds, your net result is a loss of 8 pounds. Mathematically, this is (⁻5) pounds plus (⁻3) pounds is (⁻8) pounds. If you gain 4 pounds and then gain 3 more pounds, your net result is a gain of 7 pounds. Mathematically, you have (+4) pounds plus (+3) pounds is (+7) pounds.

You might also picture both these cases with the real number line:

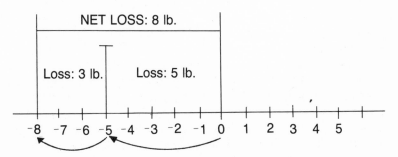

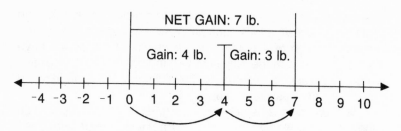

These two examples bring us to *Rule 1*: The sum of two positive numbers is a positive number. The sum of two negative numbers is a negative number.

For example, ⁻5 plus ⁻3 or (⁻5) + (⁻3) = ⁻8 and (+4) + (+3) = +7. In the second example the plus signs are usually dropped to give us the more familiar 4 + 3 = 7.

I also want to mention a case not stated in Rule 1, and that is that $0 + x = x$, for any number x.

Subtraction can now be defined in terms of signed numbers. Both 5 − 3 and 5 + ⁻3 represent the case of gaining 5 pounds and then losing 3 pounds, for a net gain of 2 pounds.

In general, a − b is the same as a + (⁻b). From here on, 7 − 5 and 7 + ⁻5 may be used interchangeably. Before going on to Rule 2, look at the following example: What is 5 − 8? The expression 5 − 8 represents a gain of 5 pounds followed by a loss of 8 pounds, for a net loss of 3 pounds. We have 5 − 8 equal to a net loss of 3, or 5 − 8 = ⁻3.

Rule 2: When adding two numbers with opposite signs the answer will be the difference between the two numbers and will have the sign of the larger.

We can picture 4 − 6 or 4 + (−6) = −2 on the number line below:

Suppose you want to calculate 2 times (⁻3) or (2) (⁻3). Two times (⁻3) is the same as (⁻3) + (⁻3), i.e., twice the quantity ⁻3, pictured below.

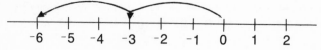

From the diagram we can see that the result is ⁻6. In the same way, (10) (⁻3) would be equal to ⁻30. This example is extended in Rule 3.

Rule 3. If *a* and *b* have opposite signs, then the product *ab* is *negative*.
Example:

$$(3) (^-4) = {}^-12$$

Now suppose you want to multiply (⁻2) times (⁻3). You will see that (⁻2)(⁻3) is actually +6. This step is one of the hardest to explain and various writers have given various explanations. I will include an explanation I found useful in the Math Without Fear program.

You know that ⁻2 is the opposite of 2. You might think of (⁻2) times (⁻3) as the opposite of (2) times (⁻3). Thus, (⁻2) (⁻3) is the opposite of ⁻6. As we see on the number line below, the opposite of ⁻6 is 6.

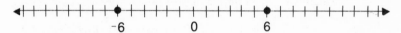

We have *Rule 4:* If *a* and *b* are both negative, then *ab* is positive.

The only case we have not considered is the following case: (+2) (+3) means we have 2 times 3 or (3 + 3), which is +6.

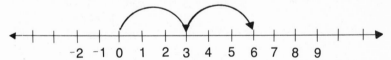

Rule 5: If *a* and *b* are both positive numbers then *ab* is also positive.

The last rule with signed numbers is fairly easy to see.

Rule 6: A number, *a*, plus its opposite, −*a*, is 0. That is, a − a = a + (−a) = 0.

To see Rule 6 look at the following example: if you gain 2 pounds and then lose 2 pounds, what is your net result? You're back where you started.

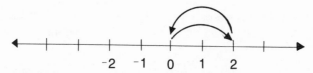

-2   -1   0   1   2

Another related property used in arithmetic is that by definition 0 times any number is always 0. For example. 0 · 5 = 0.

You may be wondering what happened to division. By definition 6 ÷ 3, or 6/3, is another symbol for 6 times 1/3. The fraction 1/3 is called the *reciprocal* of 3, i.e., it is the number so that 3 times this number equals 1. Likewise, the reciprocal of 1/8 is 8. Just as subtraction was really adding the opposite, division is really multiplying by the reciprocal. All of the rules for multiplying signed numbers extend to division. Look at the examples below:

(1) $\dfrac{-6}{-3}$ = −6 times $^-\dfrac{1}{3}$, which is +2

(negative times negative).

(2) $\dfrac{-10}{5}$ = −10 times $\dfrac{1}{5}$, which is ⁻2

(negative times positive).

DISTRIBUTIVE PROPERTY

Suppose you have the following expression: 2(1 + 5) it stands for twice the sum 1 + 5, or (1 + 5) + (1 + 5). From elementary arithmetic you know that the order in which you add these numbers does not matter. (Two apples plus 5 apples plus 1 apple is the same as 5 apples plus 1 apple plus 2 apples. In either case you get 8 apples.) This rule also holds for multiplication, that is, the order in which you multiply numbers does not matter. For example: 2 · (3 · 7) = (2 · 3) · 7. Thus,

$$1 + 5 + 1 + 5$$
$$= 1 + 1 + 5 + 5$$

But 1 + 1 is twice 1 and 5 + 5 is twice 5.
We have 1 + 1 + 5 + 5 = 2(1) + 2(5).
Conclusion: 2(1 + 5) = 2(1) + 2(5)
In general, $a(b + c) = ab + ac$.
In particular, suppose we have

$$^-3(2 + 5 - 4)$$

By the distributive law,

$$= {}^-3(2 + 5 + {}^-4)$$
$$= {}^-3(2) + ({}^-3)(5) + ({}^-3)({}^-4)$$
$$= {}^-6 {}^-15 + 12$$
$$= {}^-21 + 12$$
$$= {}^-9$$

In the above example you might note that multiplying an expression inside parentheses, such as $(2 + 5 - 4)$ by a negative such as ${}^-3$ is equivalent to multiplying each term by 3 and then changing the signs:

$${}^-3(2 + 5\ 4) = 3 \cdot 2 - 3 \cdot 5 + 3 \cdot 4$$
$$= -6 - 15 + 12 = \quad -9$$

For an excellent treatment of fractions see *Mind over Math* by Dr. Stanley Kogelman and Dr. Joseph Warren. Another excellent review of fractions can be found in *Overcoming Math Anxiety* by Sheila Tobias (see Bibliography).

APPENDIX B
SOLUTIONS

Chapter 4

(1) Let p equal Andrea's weight in pounds.
$$p + 10 = 133$$
$$p + 10 - 10 = 133 - 10$$
$$p = 123$$

(2) Let c equal the cost of eggs originally.
$$c + 8 = 89$$
$$c + 8 - 8 = 89 - 8$$
$$c = 81$$

(3) Let x equal last year's attendance.
$$124,000 = x + 15,000$$
$$124,000 - 15,000 = x + 15,000 - 15,000$$
$$109,000 = x$$

(4) Let p equal the original price of the stove.
$$275 = p - 225$$
$$275 + 225 = p - 225 + 225$$
$$500 = p$$

(5) Let m equal the number of miles from Syracuse.
$$150 + m = 295$$
$$150 - 150 + m = 295 - 150$$
$$m = 145$$

(6) Let n equal the number of children each in families No. 1 and No. 2.
$n - 1 =$ number of children in family No. 3
$2n =$ number of children in family No. 4
$n + n + (n - 1) + 2n = 14$
$n + n + n - 1 + 2n = 14$
$n + n + n + 2n - 1 = 14$
$5n - 1 = 14$
$5n - 1 + 1 = 14 + 1$
$5n = 15$
$\dfrac{5n}{5} = \dfrac{15}{5}$
$n = 3$
Conclusion: number of children in families No. 1, 2, 3, and 4 $=$ 3, 3, 2, and 6.

(7) Let h equal Harriet's age.
Edith's age $= 51$
$51 + h = 94$
$51 - 51 + h = 94 - 51$
$h = 43$

(8) $2n + 7 = 43$
$2n + 7 - 7 = 43 - 7$
$2n = 36$
$\dfrac{2n}{2} = \dfrac{36}{2}$
$n = 18$

(9) Let n equal the smaller number.
larger number $= 3n - 5$
$n + (3n - 5) = 11$
$n + 3n - 5 + 5 = 11 + 5$
$4n = 16$
$\dfrac{4n}{4} = \dfrac{16}{4}$
$n = 4$
$3n - 5 = 3(4) - 5 = 7$

(10) Let c equal the cost of clothing.

Let f equal the cost of food.
Let r equal the rent.
$f = 2c$
$r = 3f = 3\,(2c) = 6c$
$c + f + r = 540$ is equivalent to
$c + 2c + 6c = 540$
$9c = 540$
$\dfrac{9c}{9} = \dfrac{540}{9}$
$c = 60; f = 2c = 120; r = 6c = 360.$

(11) Let x equal ounces butter for icing.
$2x =$ ounces butter for cake alone
$x + 2x = 24$
$3x = 24$
$\dfrac{3x}{3} = \dfrac{24}{3}$
$x = 8$

(12) Let x equal the number of calories burned while sitting for one hour.
$3x =$ calories burned while walking
$8x + 1(3x) = 1{,}100$
$11x = 1{,}100$
$\dfrac{11x}{11} = \dfrac{1{,}100}{11}$
$x = 100$
$3x = 3(100) = 300 =$ calories burned while walking.

(13) Let x equal the length of the second piece.
$x + 5 =$ length of third piece
$7 + x + (x + 5) = 30$
$7 + x + x + 5 = 30$
$7 + 5 + 2x = 30$
$12 + 2x = 30$
$12 - 12 + 2x = 30 - 12$
$2x = 18$
$\dfrac{2x}{2} = \dfrac{18}{2}$
$x = 9$
Lengths of pieces: 7, 9, and 14 inches.

(14) Let j equal the height of Jane's hat.
$2j$ = height of Maryanne's hat
$62 + 2j = 64 + j$
$62 - 62 + 2j = 64 - 62 + j$
$2j = 2 + j$
$2j - j = 2 + j - j$
$j = 2$ = height of Jane's hat
$2j = 4$ = height of Maryanne's hat

Chapter 5

(1) $\dfrac{\text{food}}{\text{income}} = \dfrac{2{,}580}{8{,}600} = \dfrac{3}{10}$

(2) $\dfrac{3}{69} = \dfrac{5}{x}$
$3x = 345$
$x = 115¢ = \$1.15$

(3) $\dfrac{10}{1} = \dfrac{x}{5}$
$50 = x$

(4) $\dfrac{5}{3} = \dfrac{20{,}000}{x}$
$5x = 60{,}000$
$x = 12{,}000$

(5) $\dfrac{\frac{2}{3}}{2} = \dfrac{x}{7}$
$\frac{2}{3} \cdot 7 = 2x$
$\frac{14}{3} = 2x$
$\frac{1}{2} \cdot \frac{14}{3} = \frac{1}{2} \cdot 2x$
$\dfrac{7}{3} = \dfrac{14}{6} = x$

(6) $\dfrac{12}{216} = \dfrac{x}{360}$
$12 \cdot 360 = 216x$
$4{,}320 = 216x$
$\dfrac{4{,}320}{216} = x$
$20 = x$

(7) $\dfrac{1.1 \text{ yards}}{1 \text{ meter}} = \dfrac{x \text{ yards}}{10 \text{ meters}}$

$(1.1)\,10 = 1x$

$11 = x$

(8) $\dfrac{3}{6} = \dfrac{x}{10}$

$30 = 6x$

$5 = x$

(9) $\dfrac{8}{10} = \dfrac{x}{25}$

$200 = 10x$

$20 = x$

(10) $\dfrac{125}{5} = \dfrac{x}{20}$

$5x = 2,500$

$x = 500$

(11) $\dfrac{4}{1} = \dfrac{x}{7/2}$

$4 \cdot (7/2) = x$

$14 = {}^{28}/_{2} = x$

(12) $\dfrac{1/2}{5/4} = \dfrac{4}{x}$

$\tfrac{1}{2}x = (5/4) \cdot 4$

$\tfrac{1}{2}x = {}^{20}/_{4}$

$x = 2 \cdot (^{20}/_{4})$

$x = {}^{40}/_{4} = 10$

Chapter 6

MORE PRACTICAL PROBLEMS

(1) increase $= 600$

$600 = \dfrac{x}{100} \cdot 2,400$

$\dfrac{(600)\,(100)}{2,400} = x$

$25 = x$

(2) 25% of $28 = \dfrac{25}{100} \cdot 28 = 7$

sales price $= 28 - 7 = 21$

(3) discount $= 4\%$ of $26 =$

$\dfrac{4}{100} \cdot (26) = \$1.04$

(4) 12% of $x = 1{,}920$

$\dfrac{12}{100} \cdot x = 1{,}920$

$12x = 192{,}000$

$x = 16{,}000$

(5) increase $= 5¢$

5 is $x\%$ of 85

$5 = \dfrac{x}{100} \cdot 85$

$500 = 85x$

$5.88 = x$

increase is approximately 6%

Chapter 6

MORE PERCENTAGE PROBLEMS

(1) $\dfrac{2}{100} \cdot 150 = 3$

(2) $\dfrac{35}{100} \cdot 500 = 175$

(3) $\dfrac{75}{100} \cdot 120 = 90$

(4) $9 = \dfrac{x}{100} \cdot 36$

$900 = 36x$

$25 = x$

(5) $12 = \dfrac{x}{100} \cdot 80$

$1{,}200 = 80x$

$15 = x$

(6)
$$24 = \frac{x}{100} \cdot 40$$
$$2{,}400 = 40x$$
$$60 = x$$

(7)
$$15 = \frac{60}{100} \cdot x$$
$$1{,}500 = 60x$$
$$25 = x$$

(8)
$$8 = \frac{40}{100} \cdot x$$
$$800 = 40x$$
$$20 = x$$

(9)
$$36 = \frac{200}{100} \cdot x$$
$$3{,}600 = 200x$$
$$18 = x$$

(10) $5\frac{1}{2}\%$ of $600 =$
$$\frac{5\frac{1}{2}}{100} \cdot 600 =$$
$$\frac{11\frac{1}{2}}{100} \cdot 600 =$$
$$\frac{6{,}600\frac{1}{2}}{100} = \frac{3{,}300}{100} = 33$$

(11)
$$120 = \frac{x}{100} \cdot 800$$
$$12{,}000 = 800x$$
$$15 = x$$

Chapter 7

EVERYDAY PROBLEMS

(1) $A = L \cdot W$
$$80 = (W + 2)W$$

$80 = W^2 + 2W$
$0 = W^2 + 2W - 80$
$0 = (W + 10)(W - 8)$
$W = -10$ or $W = 8$
answer: $W = 8$ and $L = 10$

(2) $x^2 + 3x + 10 = 50$
$x^2 + 3x - 40 = 0$
$(x + 8)(x - 5) = 0$
$x = -8, x = 5$
answer: 5 frames

(3) $x^2 + 2x - 120 = 0$
$(x + 12)(x - 10) = 0$
$x = -12$ or $x = 10$
answer: 10 people

Chapter 8

PYTHAGOREAN THEOREM PROBLEMS

(1) $5^2 = 4^2 + a^2$
$25 = 16 + a^2$
$25 - 16 = 16 - 16 + a^2$
$9 = a^2$
$3 = a$

(2) $c^2 = 8^2 + 15^2$
$c^2 = 64 + 225$
$c^2 = 289$
$c = 17$

Chapter 8

PRACTICE PROBLEMS, AREAS

(1) $9 \cdot 9 = 81$ square centimeters

(2) $3 \cdot 12 = 36$ square feet

(3) $11 \cdot 14 = 154$ square feet

(4) Area of garden $= 12 \cdot 6 = 72$ square feet,;
72 pounds of fertilizer are needed.

(5) Area of playground $= 50 \cdot 35 = 1,750$ square yards; cost of paving playground $= \$3.50 \cdot 1750 = \$6,125.$

BIBLIOGRAPHY

Bosstick, M., and Cable, J. *Patterns in the Sand.* Beverly Hills, California: Glencoe Press, 1975.

Britton, J., and Bello, I. *Topics in Contemporary Mathematics.* New York: Harper & Row, 1975.

Carman, R., and Carman, M. *Basic Algebra.* New York: John Wiley & Sons, 1977.

Donaghey, R., and Ruddel, J. *Fundamentals of Algebra.* New York: Harcourt Brace Jovanovich, 1978.

Ernest, John. "Mathematics and Sex." *American Mathematical Monthly* 83 (October 1976).

Federer, Walter T. *Statistics and Society.* New York: Marcel Dekker, 1973.

Gallese, Liz. "A Little Calculating and a Lot of Terror Equal Math Anxiety," *Wall Street Journal*, March 13, 1978.

Gilligan, L., and Nenno, R. *Basic Algebra: A Semi-Programmed Approach.* Santa Montica, California: Goodyear, 1977.

Graham, Malcolm. *Modern Elementary Mathematics.* New York: Harcourt Brace Jovanovich, 1975.

Hackworth, R., and Howland, J. *Introduction to College Mathematics.* Philadelphia: W. B. Saunders, 1976.

Huff, Darrell. *How to Lie with Statistics.* New York: W. W. Norton, 1954.

Jacobs, Harold R. *Mathematics, a Human Endeavor.* San Francisco: W. H. Freeman, 1970.

Kogelman, S., and Warren, J. *Mind over Math.* New York: Dial Press, 1978.

Larsen, R., and Stroup, D., *Statistics in the Real World.* New York: Macmillan, 1976.

Misner, Fredric, *Mathematics: A Survey of Its Foundations.* San Francisco: Canfield Press, 1975.

Moore, David S. *Statistics, Concepts and Controversies.* San Francisco: W. H. Freeman, 1979.

Nielsen, Kaj L. *Mathematics for Practical Use.* New York: Barnes & Noble, 1962.

Rappaport, Karen D. "Sexual Roles and Mathematical Expectations." *The MATYC Journal* 12 (Fall 1978).

Rich, Barnett. *Review of Elementary Mathematics.* Schaum's Outline Series. New York: McGraw-Hill, 1977.

Rogers, J., VanDyke, J., and Barker, J. *Basic Mathematics, A Review.* Philadelphia: W. B. Saunders, 1979.

Rosenberg, R., and Risser, J. *Consumer Math and You.* New York: McGraw-Hill, 1979.

Sanders, D., Murph, A., and Eng, R. *Statistics, A Fresh Approach.* New York: McGraw-Hill, 1976.

Schockley, James E. *A Survey of General Mathematics.* New York: Holt, Rinehart & Winston, 1976.

Selby, Peter H. *Practical Algebra.* New York: John Wiley & Sons, 1974.

Sells, Lucy. "Mathematics—A Critical Filter." *Science Teacher* 45 (February 1978).

Sperling, A., and Stuart, M. *Mathematics Made Simple.* Garden City, New York: Doubleday, 1962.

Tobias, Sheila. *Overcoming Math Anxiety.* New York: W. W. Norton, 1978.

———. "Who's Afraid of Math and Why?" *Atlantic Monthly* (September 1978).

Wheeler, R. E., and Wheeler, E. R. *Mathematics: An Everyday Language.* New York: John Wiley & Sons, 1979.

Youse, Bevan. *An Introduction to Mathematics.* Boston: Allyn & Bacon, 1974.

INDEX

ABOUT THE AUTHOR

While growing up in Wilkes-Barre, Pennsylvania, Carol Gloria Crawford was always challenged by mathematics in her daily life. Her father's occupations, engineer and illusionist, sparked her interest.

She earned a B.A. summa cum laude from College Misericordia in Pennsylvania and both an M.A. and Ph.D. from Georgetown University. While a teaching fellow at Georgetown, she developed the popular Math Without Fear program. In 1973 she was elected to *Who's Who Among Students in American Universities and Colleges*. Dr. Crawford is presently an assistant professor at Le Moyne College, Syracuse, New York. *Math Without Fear* is her first book.